Price Elasticity

Schriften zu Marketing und Handel

Herausgegeben von Martin Fassnacht

Band 16

Evelyn Friedel

Price Elasticity

Research on Magnitude and Determinants

Bibliographic Information published by the Deutsche Nationalbibliothek
The Deutsche Nationalbibliothek lists this publication
in the Deutsche Nationalbibliografie; detailed bibliographic
data is available in the internet at http://dnb.d-nb.de.

Zugl.: Vallendar, Wiss. Hochsch. für Unternehmensführung, Diss., 2013

Library of Congress Cataloging-in-Publication Data
Friedel, Evelyn, 1978-
  Price elasticity : research on magnitude and determinants / Evelyn Friedel.
    pages cm. -- (Schriften zu Marketing und Handel, ISSN 1862-605X ; Band 16)
  Includes bibliographical references.
  ISBN 978-3-631-64705-9
  1. Pricing. 2. Elasticity (Economics) 3. Prices. I. Title.
  HF5416.5.F75 2014
  338.5'2--dc23
                        2014036694

D 992
ISSN 1862-605X
ISBN 978-3-631-64705-9 (Print)
E-ISBN 978-3-653-04294-8 (E-Book)
DOI 10.3726/978-3-653-04294-8
© Peter Lang GmbH
Internationaler Verlag der Wissenschaften
Frankfurt am Main 2014
All rights reserved.
PL Academic Research is an Imprint of Peter Lang GmbH.

Peter Lang – Frankfurt am Main · Bern · Bruxelles · New York ·
Oxford · Warszawa · Wien

All parts of this publication are protected by copyright. Any
utilisation outside the strict limits of the copyright law, without
the permission of the publisher, is forbidden and liable to
prosecution. This applies in particular to reproductions,
translations, microfilming, and storage and processing in
electronic retrieval systems.

This publication has been peer reviewed.

www.peterlang.com

Foreword

Price management has a high importance within the marketing field. Price is by far the most sensitive profit lever that managers can influence. Very small price changes translate into enormous changes in profit. Not only the knowledge about the magnitude of the price elasticity but also the knowledge about the determinants influencing the price reaction is essential. It is fundamental for the development of a successful marketing strategy to understand how price elasticities vary with market and product characteristics.

The work of Dr. Friedel makes a significant contribution to pricing research in two main areas: the magnitude of price elasticity and the determinants of price elasticity. To highlight is her unique dataset that stems from two data sources, academic publications and consulting projects. Her work comprises 863 price elasticities derived from 46 academic articles and 440 price elasticity cases from 62 consulting projects.

For the academic data, Dr. Friedel's work goes beyond previous meta-analytic research by analyzing price elasticities more in detail. Price elasticities are not only compared on a product level but also compared across studies on a brand and brand size level. Therefore, her research provides new meta-analytic insights on the magnitude of price elasticities. In the consulting project data, the focus is on price elasticities derived from survey data. This data extends the knowledge on price elasticities as previous price elasticity analyses stem primarily from scanner data. The use of real business data ensures the high degree of managerial relevance in her research.

Another strength of the research is the enhanced comparability of the price elasticity data by using a more consistent estimation method than in previous research. In addition, the price elasticities are both analyzed for a price decrease and a price increase, which addresses an important research gap. Determinants of price elasticities are explored beyond previously assessed variables. Another contribution of the research at hand is that determinants are analyzed overall but also for various industry settings. A broader range of products and more diverse industries are covered in her research compared to previous research, which is primarily based on fast moving consumer goods.

The research provides a detailed overview on price elasticities for a broad product spectrum covering not only business-to-consumer markets but also business-to-business markets. The focus on regular prices and long-term effects addresses another gap in research, as previous price elasticity research is primarily based on short-term effects and price promotions.

The research results of Dr. Friedel are not only relevant for the academic field but also managerial practice. I am convinced that her work will be of high interest for both academics and managers.

Prof. Dr. Martin Fassnacht
Holder of the Otto Beisheim Endowed Chair of Marketing and Commerce
WHU – Otto Beisheim School of Management

Preface

This work was accepted as a dissertation in November 2013 by the doctoral committee of WHU – Otto Beisheim School of Management. This doctoral thesis would have never been possible without the commitment of my supervisors, colleagues, friends, and family. Therefore, I would like to thank some people who have supported me in the successful completion of this dissertation.

I want to thank my first supervisor Prof. Dr. Martin Fassnacht for his support. He accompanied my academic education from the time being a student at University of Mannheim to accomplishing my dissertation at WHU – Otto Beisheim School of Management. During all these years I could rely on him as a mentor on both my academic work and my professional career.

I would like to thank my second supervisor Prof. Dr. Peter Witt for providing his timely and highly valuable second opinion.

This research was supported by my former employer Simon-Kucher & Partners, a global consulting firm specialized in strategy and marketing, through a scholarship and generous access to expert knowledge and consulting project data. Many colleagues supported the research providing insights and knowledge based on project experience. Especially, I would like to thank the company founder, Prof. Dr. Dr. h.c. mult. Hermann Simon, for his support and his feedback on the research topic. In addition, I would like to thank my former boss Dr. Klaus Hilleke and all colleagues for supporting my research project.

Finally, I want to thank my family and friends. Therefore, I would like to express my gratitude to all my friends that supported and encouraged me. Most important, I would like to thank my dad Werner Friedel and my mom Thea Friedel, as well as my sister Svenja Friedel and my brother Patrick Friedel. My family supported me in all stages of my life and contributed considerably to the successful completion of this research project. Therefore, I dedicate this work to them.

Dr. Evelyn Friedel

Contents

Foreword V

Preface VII

List of Figures XI

List of Tables XV

1 Aim of Research and Overview 1
   1.1 Relevance and Contribution of the Research 1
   1.2 Outline of the Research 7

2 Foundation of Research 9
   2.1 Conceptual and Theoretical Background 9
      2.1.1 Definition and Explanation of Price Elasticity Concept 9
      2.1.2 Price Elasticity in Relationship to Price Response Function ... 10
      2.1.3 Price Elasticity and Price Optimization 13
      2.1.4 Terminology and Related Concepts 15
   2.2 Prior Research 17
      2.2.1 Magnitude of Price Elasticity 18
      2.2.2 Determinants of Price Elasticity 19

3 Data 27
   3.1 Academic Publications 27
      3.1.1 Data Collection 27
      3.1.2 Overview of Selected Studies 29
   3.2 Consulting Projects 39
      3.2.1 Data Collection 40
      3.2.2 Overview of Selected Projects 42

4 Magnitude of Price Elasticity 48
   4.1 Insights from Academic Publications 48
      4.1.1 Overview and Analysis of Price Elasticity Data 48
      4.1.2 Analysis of Selected Product Categories 52

4.2 Insights from Consulting Projects 67
    4.2.1 Overview and Analysis of Price Elasticity Data 67
    4.2.2 Analysis of Selected Product Categories 74
4.3 Summary 82

5 Determinants of Price Elasticity 85
   5.1 Insights from Academic Publications 86
      5.1.1 Market Share 87
      5.1.2 Competition 89
      5.1.3 Premium Positioning and Quality 91
      5.1.4 Brand Ownership 93
      5.1.5 Direction of Price Change 95
      5.1.6 Customer Characteristics 97
   5.2 Insights from Consulting Projects 99
      5.2.1 Selection of Determinants 100
      5.2.2 Coding of Determinants 102
      5.2.3 Expected Relationship of Determinants with Price Elasticity ... 110
      5.2.4 Analysis of Determinants 118
          5.2.4.1 Analysis of Determinants - Average Price Elasticity 119
          5.2.4.2 Analysis of Determinants - Price Elasticity for a Price Decrease 124
          5.2.4.3 Analysis of Determinants - Price Elasticity for a Price Increase 126
          5.2.4.4 Synthesis of Results 128
      5.2.5 Industry Specific Aspects 130
   5.3 Summary 138

6 Conclusions 144
   6.1 Key Findings 144
   6.2 Implications for Research 150
   6.3 Implications for Management 153

Appendix 156

References 169

List of Figures

Figure 1-1: Price Elasticity Representing the Core of Pricing 1

Figure 2-1: Price Response Functions and Price Elasticities 11

Figure 2-2: Frequency Distribution of Price Elasticities – 1961-2004 18

Figure 4-1: Frequency Distribution of Price Elasticities – Academic Data Set 49

Figure 4-2: Frequency Distribution of Price Elasticities – Laundry Detergent 52

Figure 4-3: Frequency Distribution of Price Elasticities – Ketchup 53

Figure 4-4: Frequency Distribution of Price Elasticities – Bathroom Tissues 55

Figure 4-5: Frequency Distribution of Price Elasticities – Automotive Hard Parts 56

Figure 4-6: Frequency Distribution of Price Elasticities – Yogurt 57

Figure 4-7: Frequency Distribution of Price Elasticities – Margarine 59

Figure 4-8: Frequency Distribution of Price Elasticities – Peanut Butter 60

Figure 4-9: Frequency Distribution of Price Elasticities – Coffee 61

Figure 4-10: Frequency Distribution of Price Elasticities – Coffee: Preference Measurement 62

Figure 4-11: Frequency Distribution of Price Elasticities – Shampoo 63

Figure 4-12: Frequency Distribution of Price Elasticities – Tuna 64

Figure 4-13: Frequency Distribution of Price Elasticities –
Saltine Crackers..65

Figure 4-15: Frequency Distribution of Price Elasticities –
Price Decrease of 10%..69

Figure 4-16: Frequency Distribution of Price Elasticities –
Price Increase of 10%...70

Figure 4-17: Frequency Distribution of Price Elasticities –
Price Decrease of 10 % (all Data)...72

Figure 4-18: Frequency Distribution of Price Elasticities –
Price Increase of 10% (all Data)...73

Figure 4-19: Frequency Distribution of Price Elasticities –
Automotive 75

Figure 4-21: Frequency Distribution of Price Elasticities –
Industrial Goods..77

Figure 4-22: Frequency Distribution of Price Elasticities –
Away from Home Food..78

Figure 4-23: Frequency Distribution of Price Elasticities –
Logistics..79

Figure 4-24: Frequency Distribution of Price Elasticities –
Consumer Durables 80

Figure 4-25: Frequency Distribution of Price Elasticities –
Pharmaceuticals and Medical Technology...81

Figure 5-1: Categories of Determinants Identified in Academic Publications.........87

Figure 5-2: Research Framework for Determinants of Price Elasticity 101

Figure 5-3: Results of Regression Analysis –
Determinants Analyzed with Average Price Elasticity......................120

Figure 5-4: Results of Regression Analysis –
Determinants Analyzed with Price Elasticities
for a Price Decrease of 10% 125

List of Figures

Figure 5-5:   Results of Regression Analysis –
              Determinants Analyzed with Price Elasticities
              for a Price Increase of 10% 127

Figure 5-6:   Overview of Effects of Determinants on
              Magnitude of Price Elasticity 129

Figure 5-7:   Stepwise Regression Analysis –
              Automotive 132

Figure 5-8:   Stepwise Regression Analysis –
              Fast Moving Consumer Goods 133

Figure 5-9:   Stepwise Regression Analysis –
              Industrial Goods 134

Figure 5-10: Stepwise Regression Analysis –
              Away from Home Food 135

Figure 5-11: Stepwise Regression Analysis –
              Logistics 136

Figure 5-12: Stepwise Regression Analysis –
              Consumer Durables 137

Figure 5-13: Stepwise Regression Analysis –
              Pharmaceuticals and Medical Technology 137

Figure 5-14: Regression Analysis – Key Determinants 142

List of Tables

Table 2-1: Research Methodology: Effects on Magnitude of Price Elasticity 20

Table 2-2: Market Characteristics: Effects on Magnitude of Price Elasticity 24

Table 3-1: Selected Academic Studies on Price Elasticities 30

Table 3-2: Selected Consulting Projects on Price Elasticities 43

Table 4-1: Price Elasticities by Product Category in Academic Data Set 50

Table 4-2: Price Elasticities by Product Category and Study in Academic Data Set 51

Table 4-3: Price Elasticities of Selected Ketchup Brand Sizes 54

Table 4-4: Price Elasticities of Selected Yogurt Brands 58

Table 4-5: Price Elasticities of Dannon Yogurt 8-ounce Size 58

Table 4-6: Descriptive Values for Selected Product Categories 66

Table 4-7: Price Elasticities by Product Category 74

Table 4-8: Descriptive Values for Selected Product Categories 82

Table 4-9: Comparison of Descriptive Values for Selected Data Sources 83

Table 4-10: Comparison of Descriptive Values for a Price Decrease vs. a Price Increase 83

Table 5-1: Determinant – Market Share 88

Table 5-2: Determinant – Competition 90

Table 5-3: Determinant – Premium Positioning and Quality 92

Table 5-4: Determinant – Private Label vs. Brand 94

Table 5-5: Determinant – Direction of Price Change 96

Table 5-6: Determinant – Direction of Price Change – Study Example 96

Table 5-7: Determinant – Customer Characteristics 97

Table 5-8: Overview of Variables for Regression Analyses 119

Table 6-1:	Mean Price Elasticities – Academic Data	146
Table 6-2:	Mean Price Elasticities – Consulting Project Data	147
Table 6-3:	Categories of Determinants Identified in Academic Publications	148
Table 6-4:	Determinants of Price Elasticity – Consulting Project Data Product and Market Characteristics	149
Table 6-5:	Determinants of Price Elasticity – Consulting Project Data Research Methodology	150

1 Aim of Research and Overview

1.1 Relevance and Contribution of the Research

Price management has a high importance within the marketing field (Gijsbrechts 1993; Leone et al. 2012; Monroe 2003; Simon/Fassnacht 2009). The price indicates the ratio of "the quantities of money (or goods and services) needed to acquire a given quantity of goods or services" Monroe (2003, p. 5). The price response function and the price elasticity are the essential elements of pricing (Simon/Fassnacht 2009, p. 13). This view is supported by Kim/Srinivasan/Wilcox (1999, p. 173) who consider the price elasticity to be the most fundamental economic concept of pricing.

Figure 1-1 points out the central role of the price elasticity and its systematic context. The price and other determinants, such as product characteristics or the competitive environment, determine the sales volume which in turn impacts the revenue and the costs and thus in the end determines the profit. In order to make profitable pricing decisions, knowledge about customers' price reaction is indispensable (Simon/Dolan 1997, p. 59).

Figure 1-1: Price Elasticity Representing the Core of Pricing

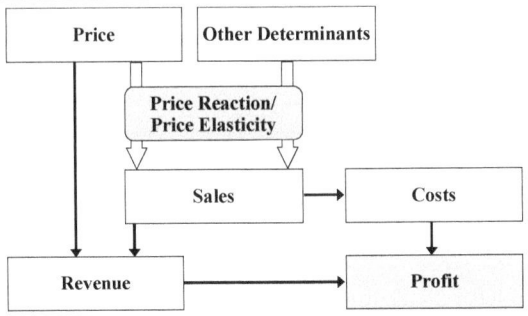

Source: adapted from Simon/Dolan 1997, p. 59

The most important key figure for a pricing decision is the price elasticity of demand (Kucher 1985, p. 193). The price elasticity indicates how sensitively consumers react to price changes. The price response function provides essential information for the pricing decision. Condensing this information into a meaningful measure, such as the price elasticity of demand, enables managers to compare products on the basis of a key figure. In combination with the information on costs, profit-optimal prices can be calculated.

Furthermore, the price elasticity is a particularly suitable key figure as it is a dimension-free measure and thus enables a comparison across brands, product categories, customers and markets (Sivakumar 2001, p. 1; Tellis 1988). Price elasticities are the most prevalent measure of price responsiveness in the marketing and economics literature (Mulhern/Williams/Leone 1998, p. 433). In addition, it is helpful to understand market structures and to develop marketing strategies (Kamakura/Russell 1989, Sivakumar 2001). Consequently, obtaining accurate estimates of price elasticities is crucial (Sivakumar 2001, p. 1).

Price is a fundamental profit-driver (Herrmann 2003) and has more influence on profit (ceteris paribus) than costs and sales as shown by Simon/Fassnacht (2009, p. 3). Increasing the price by only 2% under the assumption that sales do not decrease and costs are not affected, leads to substantial increases in profit. Simon/Fassnacht (2009, p. 5) calculated the impact of a 2% price increase on profit for 29 major companies in Germany and found an increase in profit of 3% to 184%, with a median of 23%. This shows the enormous leverage effect of the price. "Pricing right is the fastest and most effective way to grow profits" (Baker/Marn/Zawada 2010, p. 3). It is by far the most sensitive profit lever that managers can influence; very small price changes translate into enormous changes in profit (Baker/Marn/Zawada 2010, p. xiii). However, usually price increases affect demand, thus a 10% price increase will most likely lead to lower sales. In order to know how the price increase affects profit, one needs to know the impact on sales to assess the overall profitability of the price change. This is where the knowledge of the price elasticity of demand is essential to assess the price sensitivity of customers.

As mentioned, the price has a stronger impact on sales than most other factors. The impact of a price change on sales has been proven to be 10 to 22 times higher than the impact of advertising (Hanssens/Parson/Schultz 2001; Sethuraman/Tellis/Briesch 2011) and about 8 times higher than the impact of the sales force on sales given the same percentage change for all marketing variables (Albers/Mantrala/Sridhar 2008; Albers/Mantrala/Sridhar 2010; Bijmolt/Van Heerde/Pieters 2005). It is also notable, that advertising elasticities, i.e. the percentage increase in sales for a 1% increase in advertising, decline over time (Sethuraman/Tellis/Briesch 2011, p. 469). Given the relatively low responsiveness of consumers to advertising, Sethuraman/Tellis/Briesch (2011, pp. 467; 469) calculate the optimal advertising to sales ratio (Sethuraman/Tellis 1991) and find that a 30% increase in advertising is needed to offset a 1% price cut.

Not only the knowledge about the magnitude of the price elasticity but also the knowledge about the determinants influencing the price reaction is essential. It is fundamental for the development of a successful marketing strategy to understand how price elasticities vary with market and product characteristics (Bijmolt/Van Heerde/Pieters 2005, p. 141). Understanding the magnitude and the determinants of price elasticity is critical for managers in order to enhance profitability. Previous research results on price elasticities are partly contradictory and generalizations are limited (Tellis 1988; Bijmolt/Van Heerde/Pieters 2005). In addition, there is a gap between academic and managerial knowledge as many managers are not able to define the term price elasticity precisely and are not able to quantify the effect (Simon/Fassnacht 2009, p. 10). A better understanding of the price elasticity concept could enhance managerial decision making. Although price management is one of the most critical functions, it remains one of the most misunderstood and undermanaged functions at many companies (Baker/Marn/Zawada 2010, p. xiii). "Pricing is managers' biggest marketing headache" (Dolan 1995, p. 174). It is considered by managers to be a difficult area in which to set objectives and measure results; in pricing they feel the most pressure to perform and the least certainty that they are performing well (Dolan 1995, p. 174).

Given the importance of price elasticities, a large number of academic studies has been published in recent years, an overview of selected studies can be found in the meta-analysis of Bijmolt/Van Heerde/Pieters (2005, p. 143 f.) and chapter 3.1.2. The majority of these studies are econometric studies. Even though econometric price elasticity calculations on the basis of market data are considered to have a high validity (Ben-Akiva et al. 1994), they also have several drawbacks. One essential shortcoming is the fact that this methodology cannot be used for new products (Simon/Dolan 1997, p. 89). In addition, the price variation to determine the elasticity is usually rather low (Bemmaor 1984; Brodie/de Kluyver 1984; Bucklin/Russell/Srinivasan 1998) and if there is not enough variation in price, the elasticities can be hard to estimate accurately (Cooper 1988, p. 721). Econometric methods dominate the academic literature but this does not reflect their usage in the managerial world. In this setting, customer surveys, e.g. using conjoint analyses, and expert judgments play a more important role (Hensel-Börner/Sattler 2000, p. 705; Lauszus/Ebel 2000, p. 839; Weiber/Rosendahl 1997, p. 107; Simon/Fassnacht 2009, p. 105). It is estimated that each year more than 1,000 conjoint analyses are conducted in the business world (Hensel-Börner/Sattler 2000, p. 706). A significant advantage of this method is the applicability for new products (Simon/Fassnacht 2009, p. 105). Managers can estimate price response functions and price elasticities, and this provides valuable information to determine profit-optimal prices when launching a product; especially for new innovative products when no market data is available. In addition, managers can test customers' reactions towards price changes before a relaunch of a product or for example a facelift in the automobile industry. Another advantage of this data source is the high degree of managerial relevance. Using surveys can also overcome the limitation of market data regarding the low price variation as a broader price range can be explored in surveys.

To date, no meta-analysis the author is aware of has examined survey data on price elasticities. In the research directions, the most recent meta-analysis on price elasticities asks for further research beyond the range of the current meta-analytic design that uses primarily real purchase scanner data (Bijmolt/Van Heerde/Pieters

2005). One way the author addresses this in the research at hand is by taking the research beyond scanner data and reflecting the managerial relevance by utilizing survey data that stem from consulting projects. Overall, researchers seem to underestimate hypothetical approaches regarding their value in guiding managerial decision making (Miller et al. 2011, p. 182).

The intense usage of scanner data in academic research implies that the data studied consists primarily of fast moving consumer goods. In the meta-analysis of Bijmolt/Van Heerde/Pieters (2005), 98% of all price elasticities stem from fast moving consumer goods and are divided into groceries with high and low stockpiling propensities, the remaining 2% stem from durables. Given this categorization, another area to extend the scope of knowledge on price elasticities is to analyze product categories more in detail. A further exploration of which products are actually summarized under the grocery category and the analysis of their respective price elasticities will lead to additional insights. Even within one product category, such as frequently purchased packaged goods, a range of average price elasticities for each product from -0.63 for blended butter/margarine to -4.34 for jam is measured (Danaher/Brodie 2000, p. 922). An aim of this study is to gain additional knowledge on price elasticities by analyzing narrower product categories, by comparing price elasticities across studies and by composing price elasticity distributions across studies. This kind of research is currently lacking. Taking the yogurt category as an example, there are many studies with this research subject (e.g., Besanko/Gupta/Jain 1998; Chintagunta 1992; Chintagunta 1993; Kim/Allenby/Rossi 2002; Van Heerde/Gupta/Wittink 2003; Villas-Boas/Winer 1999) but no systematic overview exists on the distribution pattern of the price elasticities or the descriptive values across studies; also no analysis on the brand level across studies exists in the academic literature. This research aims to provide this additional information currently lacking in research.

It is also pointed out by researchers that the interpretation and comparison of price elasticities is difficult due to the absence of comparable data (Gordon/Goldfarb/Li 2013, p. 5). Even given the same data, the price elasticity estimated by the researchers can be quite different. Song/Chintagunta (2007, p. 609), who use the same data source as Bell/Chiang/Padmanabhan (1999), notice that the price elasticities estimated by Bell/Chiang/Padmanabhan (1999) (ranging from -4.66 to -5.66) are "much larger" than their own estimates (ranging from -1.47 to -2.00) and attribute the "dramatic differences in the results between the studies" to the differences in the research methodology. The currently available meta-analyses (primarily Bijmolt/Van Heerde/Pieters 2005; Tellis 1988) compare price elasticities that are calculated via a broad variety of methodologies. The author of this research will build a more comparable data set based on consulting projects because a consistent methodology is used for calculating price elasticities.

Given that fast moving consumer goods are the backbone of previous research, another way to extend the scope of knowledge on price elasticities is to expand the range of products beyond fast moving consumer goods. The data derived from the consulting projects will broaden the database beyond the typical research setting and thus provide information of product categories previously not integrated in the research stream. This will make industry specific analyses accessible and provide an opportunity to compare price elasticities across products and diverse industries. Currently comparative assessments are rather limited due to the lack of data.

In addition, business-to-business data on price elasticities from the consulting data will be included in the database and therefore address another research gap. A citation and profiling analysis in the most important marketing and business journals identifies business-to-business pricing as an area that has been underresearched in the 30 years from 1980 to 2010 (Leone et al. 2012, p. 1023).

In order to identify current research activities in the pricing field, Roth (2010, p. 170) analyzed the four leading marketing journals (Journal of Marketing, Journal of Consumer Research, Journal of Marketing Research and Marketing Science) as well as two additional journals that cover pricing research (Journal of Product and Brand Management, Marketing Zeitschrift für Forschung und Praxis). Articles published between 2004 and 2008 were analyzed and categorized into 26 pricing topics. The topics were then ranked according to the number of articles addressing this topic to indicate the relevance and topicality of pricing research. The top three topics are price perception, price variation and pricing in different industries. The research at hand is therefore positioned among highly relevant and current research, addressing price variation and analyzing price changes in a broad array of industries.

In contrast to the major research stream which focuses on short term price changes, especially price promotions (e.g., Anderson/Song 2004; Chandran/Morwitz 2006; Lu/Moorthy 2007), this research will examine long-term price changes to expand the knowledge. In the most recent meta-analysis on price-elasticities (Bijmolt/Van Heerde/Pieters 2005), 95% of cases examine a short-term duration of effect and only 5% a long-term effect, which illustrates the massive dominance of short-term effects being analyzed in the academic literature. Also looking at the definition of the price, it becomes apparent that only in 1% of all cases the price is defined as the regular price, which is the base price in non-promotional conditions. The other two price definitions are promotional price and actual prices. Fluctuations in actual prices reflect both regular price changes and temporary price discounts and a distinction cannot be made due to lack of more detailed data. Due to the limited number of observations on regular prices, some analyses such as temporal effects are only interpreted for actual and promotional price elasticities (Bijmolt/Van Heerde/Pieters 2005, pp. 150, 158). Lodish/Mela (2008) criticize the short-term perspective and attribute the focus on price promotions and short-term effects to the relatively easy measurability due to the scanner technology. In this research, the author will focus on regular prices and the long-term effect of price changes addressing the current lack of information in this area.

Looking at the determinants of price elasticity previously studied, it becomes apparent that the majority of determinants are related to the research methodology, which comprises primarily data and model characteristics (Bijmolt/Van Heerde/Pieters 2005, Tellis 1988, cf. chapter 2.2.2). Market and product characteristics are also explored but to a lesser degree and most determinants cannot be influenced by the management, e.g. inflation rate, household income. Therefore, the goal of this research is to identify and analyze determinants that are not yet explored in the previous meta-analyses and also ones that have a higher degree of managerial relevance.

The aim of the research is to enhance the knowledge about customers' price reactions and thus to yield critical information for designing professional pricing strategies. This addresses a currently identified research need since a recent analysis of pricing research identified the area of how to strategically manage pricing as underresearched (Leone et al. 2012, p. 1023). It is not only academic curiosity on consumers' price sensitivity but also managerial concern about appropriate price levels that drives the research on price elasticities (Tellis 1988, p. 331).

Reflecting the academic and managerial need, the objective of this research is to gain a more comprehensive understanding in two main areas, the magnitude of price elasticity and the determinants of price elasticity.

- The **first objective** is to expand the knowledge on the magnitude of price elasticity.

- The **second objective** is to expand the knowledge on determinants of price elasticity.

Given the scientific and managerial relevance of this topic and addressing the research gaps identified, the knowledge expansion will be aimed for in the following ways:

- Two data sources are examined: first, a refined set of the academic data used in previous meta-analyses and second, data derived from consulting projects that comprise primary research on price response by means of expert interviews or customer surveys. A key advantage of this additional data source is the high degree of managerial relevance and its potential use for new products before market data is available.

- The academic literature serves as the foundation to go deeper into the analysis. Specifically, the price elasticities will be explored in further details for various

Chapter 1: Aim of Research and Overview

product categories and compared across studies. In addition, analysis will show which determinants have been explicitly studied in previous research and what kind of implications can be drawn for further determinants beyond those studied to date.

- Analyzing survey data will expand the knowledge beyond primarily scanner data based research results. New data sources are utilized to enhance the understanding of price elasticities. Survey data has a high degree of managerial relevance and it seems that research has so far underestimated the potential of this data in assessing price elasticities.

- The additional data source will expand the currently available academic data set on price elasticities beyond primarily fast moving consumer goods. Looking at price elasticities derived from actual consulting projects will cover a broader range of products and diverse industry settings. Not only the business-to-consumer market but also the business-to-business market will be examined in this analysis thus a previously underresearched area is addressed.

- This research will provide access to product and industry specific analyses. Creating new data sets based on the academic data and the consulting projects allows detailed comparisons of the magnitude and the determinants of price elasticities across academic studies and consulting projects as well as across products and industries.

- The focus on regular prices and long-term price changes is another aspect to address the current lack of knowledge in this area, as current price elasticity research is primarily based on promotional price changes and short-term effects.

1.2 Outline of the Research

This work is structured in six chapters. In the first chapter the author described the motivation and aim of the research and demonstrated the relevance and contribution of the research. Following this chapter, the foundation of the research is laid (chapter 2). This includes the conceptual and theoretical background (chapter 2.1) explaining the price elasticity concept and its relationship to price response functions, price optimization and terminological aspects; as well as an overview of prior research results (chapter 2.2) on the magnitude and determinants of price elasticity.

In the third chapter the author provides information concerning the data on which the research is based on. Consecutively, data sets are built based on academic publications (chapter 3.1) and consulting projects (chapter 3.2). For both data sources, the author describes the data collection procedure given the selection criteria and provides an overview of the selected studies.

Provided this foundation, the magnitude of price elasticity is analyzed in chapter 4. First, the academic data set is assessed (chapter 4.1) and insights on the overall data and selected product categories as well as brands are presented. Second, the consulting project data (chapter 4.2) is assessed and the insights on the overall data as well as selected product categories are presented. Subsequently, a summary of the findings on the magnitude of price elasticity is provided (chapter 4.3).

To gain insights on the determinants of price elasticity (chapter 5) the research approach is explained. First, the academic publications are analyzed, specifically looking at determinants previously studied and additional information on determinants that the studies indirectly provided (chapter 5.1). In the next part, the consulting projects are addressed (chapter 5.2). The selection and coding of determinants is described. Then the findings, based on the analysis of the consulting project data, are presented and industry specific aspects are addressed. The findings on the determinants of price elasticity are summarized in the last part of this section (chapter 5.3)

Chapter 6 concludes with the key findings of the research (chapter 6.1). The implications for research including the contribution of this work as well as limitations of the research and future research directions are presented (chapter 6.2). The last part (chapter 6.3) points out the implications of the findings for management.

2 Foundation of Research

2.1 Conceptual and Theoretical Background

2.1.1 Definition and Explanation of Price Elasticity Concept

The most common measure of the impact of price on demand is price elasticity. In general, elasticity measures the relation of a relative change of one variable to the relative change of another variable. Specifically, price elasticity of demand is defined as the percentage change in quantity relative to the percentage change in price (Simon/Fassnacht 2009, p. 95; Diller 2008, p. 75; Monroe 2003, p. 32).

$$price\ elasticity = \frac{percentage\ change\ in\ demand}{percentage\ change\ in\ price} \quad (2.1)$$

Price elasticity therefore shows the percentage change in demand, if the price is changed by 1 % (Diller 2008, p. 75; Frank 2008, p. 112). If a price decrease of 10% causes an increase in demand of 30%, the price elasticity is -3. The negative value of the price elasticity is due to the inverse relationship of the change in demand and price. The price change in the example is -10% (numerator) and the demand change is +30% (denominator), which leads to a negative value of -3 for the price elasticity. The percentage change in demand is three times as high as the percentage change in price.

If the price change is not infinitely small, then the price elasticity is called arc elasticity (Simon/Fassnacht 2009, p. 95). Arc elasticities are commonly used in managerial practice. Frank (2008, p. 112) states „When interpreting actual demand data, it is often useful to have a more general definition of price elasticity that can accommodate cases in which the observed change in price does not happen to be 1 percent." This highlights the managerial relevance and usefulness of the arc elasticity that can accommodate larger prices changes, e.g. of 10%, than the infinitely small price changes in the mathematical example.

Strictly speaking, the price elasticity ε is defined as point elasticity, i.e. an observation of infinitely small changes. Therefore looking at a price response function of $q = q(p)$, with sales volume q and price p, the mathematical expression is:

$$\varepsilon = \frac{\frac{\partial q(p)}{q}}{\frac{\partial p}{p}} = \frac{\partial q(p)}{\partial p} \cdot \frac{p}{q} \quad (2.2)$$

$\frac{\partial q(p)}{\partial p}$ is the derivative of a general price response function. In the next chapter, the relationship of the price elasticity with the price response function will be further illustrated.

The price elasticity is generally negative since price changes move in the opposite direction from changes in quantity demanded. If the percentage change in quantity is larger than the percentage change in price, i.e. $\varepsilon < -1$ or $|\varepsilon| > 1$, the demand is called elastic. The demand is called inelastic if the percentage change in quantity is smaller than the percentage change in price, i.e. $\varepsilon > -1$ or $|\varepsilon| < 1$ (Diller 2008, p. 75). The demand is called unit elastic with respect to price if $\varepsilon = -1$ (Frank 2008, p. 111 f.). If $\varepsilon = 0$, the demand is perfectly inelastic since it does not react to price changes (Frank 2008, p. 114).

Even if in classical microeconomic theory the price elasticity is always negative since the "law of demand"(Marshall 1890) says there is an inverse relationship of the price and demand (Frank 2008, p. 112); there are also some cases in which the elasticity can be positive (Simon/Fassnacht 2009, p. 64; Bijmolt/Van Heerde/Pieters 2005, p. 145). Sometimes the occurrence of positive price elasticities can be attributed to the Veblen-effect (Veblen 1899). A price increase of a good can increase demand due to an increased preference of consumers to buy the product. Conspicuous consumption and status-seeking are the factors causing the Veblen-effect which can be found for example for some luxury goods. Another explanation for positive price elasticities is the Giffen paradoxon mentioned by Marshall (1895, p. 208) in later editions of his classical textbook "principles of economics" which stands in conflict to the law of demand (Stigler 1947, p. 152). The Giffen good is a special kind of inferior product. When its price rises, demand rises due to budget constraints and more of the inferior good is consumed. The example provided by Marshall is the following: if the price of bread rises for a poor family, they are forced to restrict their consumption of meat and other more expensive bakery goods, and for bread being still the cheapest food they can get, the family will consume more and not less of it (Marshall 1895, p. 208). Even though Giffen goods are theoretically possible, they are rarely observed; and if the effect is observed, it is attributed to the behavior of impoverished people and not seen as the feature of a specific product (Jensen/Miller 2008). "Giffen behavior" could be found for extremely poor households in China with respect to the consumption of rice (Jensen/Miller 2008). Even given the possibility of positive price elasticities, the price elasticity is in general negative; Bijmolt/Van Heerde/Pieters (2005) compiled a database of 1851 price elasticities and 97.8 % of the observed values are negative.

2.1.2 Price Elasticity in Relationship to Price Response Function

Looking at a price response function, usually different price elasticities result at various price points. Therefore, it is necessary to determine a point on the price response function, at which the price elasticity calculation is valid. The type of price

Chapter 2: Foundation of Research

response function is also influencing the price elasticity as shown in figure 2-1. For linear price response functions higher prices automatically lead to greater price elasticities. The multiplicative function has a constant price elasticity which is also called isoelasticity. For both the attraction as well as the Gutenberg model, the price elasticity is nonlinear. Figure 2-1 illustrates the relationship between price response functions and price elasticities.

Figure 2-1: Price Response Functions and Price Elasticities

Model	Dependent Variable	Mathematical Form
Linear	q_i or m_i	$a - bp_i + c\bar{p}$
Multiplicative	q_i or m_i	$a(p_i/\bar{p})^b$
Attraction	m_i	$a_0 + a_i p_i^{b_i} / \sum_i a_i p_i^{b_i}$
Gutenberg	q_i or m_i	$a - bp_i + c_1 \sinh(c_2(\bar{p} - p_i))$

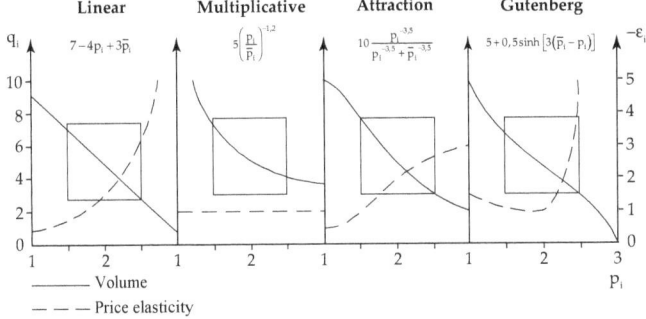

——— Volume
– – – Price elasticity

q = sales or m = market share
p = own price
$\bar{p}$ = average competitive price
Source: adapted from Simon/Fassnacht 2009, p. 100

Figure 2-1 also shows that the functions are rather similar within a certain price interval which is indicated by the squares. If prices vary only within such a narrow interval, it is almost impossible to distinguish empirically between the different models (Simon 1989, p. 21). Empirical evaluations to determine which of the models fit best are inconclusive (Brodie/de Kluyver 1984; Kucher 1985; Leeflang/Wittink 2000, Ghosh/Neslin/Shoemaker 1984). The multiplicative and the attraction model are most often used for academic research on price elasticities (Bijmolt/Van Heerde/Pieters 2005, p. 147). Simon/Fassnacht (2009, p. 100) consider the multiplicative model as less robust than the linear model; in addition they provide reasoning for the argument that price elasticities tend to be underestimated in empirical research using the multiplicative model.

One possibility to address the issue of different price elasticities at various price points is to calculate price elasticities for different prices, e.g. for a low price, for a medium-high price and for a high price. Another possibility is to determine a price that characterizes the price elasticity best, e.g. the most common price or the average price (Kucher 1985, p. 199).

Even using the identical price point, the price elasticity is often influenced by the magnitude and the direction of the price change. Gabor (1988, p. 18) states that "...there is no such thing as *the* elasticity of demand for a good, since even if we start from a given price, elasticity will tend to vary with the magnitude of the price change". Therefore, the circumstance that the magnitude of price elasticity is influenced by these factors has to be taken into account when determining how to calculate price elasticities. This is especially significant when generating a database to compare price elasticities.

As mentioned, there are rather diverse ways to calculate price elasticities and the specific calculations are not specified in most studies. Looking at the data included in the meta-analysis of Bijmolt/Van Heerde/Pieters (2005) it becomes apparent that there are substantial differences in the price spread used to calculate the price elasticities. Even within the same study, not the same price cut is applied to all products. The coefficient of variation ranges from 1% to 3% (Brodie/de Kluyver 1984), to a coefficient of variation of 5.3% (Bemmaor 1984) to fixed price changes of 10% for every product (Moon/Russell/Duvvuri 2006).

Some price elasticities are based on sales (Kadiyali/Chintagunta/Vilcassim 2000), others on market share (Kim/Rossi 1994). The direction of price change is often not specified. A single study o which the author is aware displays explicitly separate values for price elasticities based on a price increase and a price reduction (Moon/Russell/Duvvuri 2006). Other examples are price elasticities with and without category choice (Sivakumar 2001) or promotional vs. regular price elasticities (Guadagni/Little 1983).

In general, different model specifications within studies lead to multiple estimates, e.g. 3 brands and 10 or 13 different models result in 30 and 39 price elasticity estimates respectively for these three brands (Ailawadi/Gedenk/Neslin 1999; Chintagunta 2001). The use of multiple models is a common practice and thus increases the available number of price elasticities (e.g., Kim 1995, Kim/Allenby/Rossi 2002; Murthi/ Srinivasan 1999).

In the research at hand, these aspects are addressed and a database is derived from the consulting project data that is more similar, for example, in terms of the price anchor (the starting point of the price change), the direction of the price change, the magnitude and the temporal pattern of the price change. A data set with price elasticities that are generated using the same methodology allows a direct comparison of price elasticities and distinguishes the study from previous research. The interpretation and comparison of price elasticity data is difficult in the absence of comparable data and methodologies, therefore it is important to compare "apples to apples" and not "apples to pears" (Gordon/Goldfarb/Li 2013, p. 5).

2.1.3 Price Elasticity and Price Optimization

The price elasticity also plays an important role in determining profit-optimal prices. This can be illustrated via classical price theory using microeconomic concepts. The monopolistic case is used as an illustration for the relationship between the optimal price and the price elasticity. Standard optimization theory says that an optimization problem can be solved by taking a derivative and setting it equal to zero. Specifically, the descriptions of Simon/Fassnacht (2009, pp. 207-209) and Simon (1989, p. 54 f.) serve as a guideline.

First, the optimality condition for a general price response function $q(p)$ is derived. Revenue R is defined as price p times sales volume q so that the following revenue function is obtained.

$$R = p \cdot q(p) \quad (2.3)$$

Profit P is equal to revenue R minus cost C, thus the profit function is:

$$P = R - C = p \cdot q(p) - C[q(p)] \quad (2.4)$$

In the next step, the goal is to maximize the profit function with regard to the price p. Optimality conditions are best defined via the concepts of marginal revenue and marginal cost. Consequently, the profit function is derived with respect to the price p.

$$\frac{\partial P}{\partial p} = q(p) + p \frac{\partial q}{\partial p} - \frac{\partial C}{\partial q} \frac{\partial q}{\partial p} \quad (2.5)$$

At the optimal price p^* the derivative of the function equals zero.

$$\frac{\partial P}{\partial p} = q(p^*) + p^* \frac{\partial q}{\partial p} - \frac{\partial C}{\partial q}\frac{\partial q}{\partial p} = 0 \quad (2.6)$$

A different notation for the same condition is:

$$q(p^*) + p^* \frac{\partial q}{\partial p} = \frac{\partial C}{\partial q}\frac{\partial q}{\partial p} \quad (2.7)$$

The left hand side of equation (2.7) is marginal revenue and the right hand side is marginal cost, both with respect to price. This equation simply states that at the optimal price p^* marginal revenue equals marginal cost. The positive effect of a price increase on unit contribution is exactly cancelled out by the negative effect of the price increase on sales and vice versa. A divergence from this condition results in decreased profit since either the cost increases more than the revenue (deviation of the price below the optimal price) or the revenue decreases more than the cost (deviation above the optimal price).

No fixed term appears in the optimality condition (2.7). The optimal price is therefore independent of fixed cost and fixed cost should not be taken into account in price settings. This is an important result since many practitioners believe that fixed cost influence the profit-optimal price, but as illustrated taking fixed cost into account is logically not coherent.

The optimality condition (2.7) can be simplified by using the price elasticity term as defined previously:

$$\varepsilon = \frac{\partial q(p)}{\partial p} \frac{p}{q} \quad (2.2)$$

Multiplying equation (2.7) by p^*/q, substituting ε for the price elasticity term and solving for p^*, results in:

$$p^* = \frac{\varepsilon}{1+\varepsilon} C' \quad (2.8)$$

$C' = \partial C / \partial q$ represents the marginal cost with respect to quantity. Equation 2.8 is called the Amoroso-Robinson Relation. The optimal price is expressed as an elasticity-dependent mark-up factor on the marginal cost C'. It should be pointed out that the elasticity ε has a negative sign.

This equation generally does not represent a solution for p^* but is just a transformation of the necessary optimality condition "marginal revenue = marginal cost" (2.7) because both ε and C' may depend on p^*. It is a so-called fixed point equation. The relation, however, provides a convenient way to calculate the optimal price in the case of a constant-elasticity price response function, e.g. multiplicative form (cf. chapter 2.1.2).

Looking at the Amoroso-Robinson Relation, it becomes apparent that the optimal price p^* has to be in the range of $\varepsilon < -1$. The optimal price is thus greater than the revenue-maximizing price at which $\varepsilon = -1$. Independent of the value of the marginal cost, it is profit-maximizing to raise the price if the absolute price elasticity is smaller than one. If a price increase of 10% lowers demand by 5% ($\varepsilon = -0.5$), it makes sense to raise the price independent of the marginal cost. The percentage change in quantity is less than the percentage change in price, thus, when the price is raised, the profit rises.

Formula (2.8) cannot be used to derive the optimal price when marginal cost equals zero as this can be the case with certain software or services. Generally, the higher the absolute price elasticity, the lower is the profit-optimal price. This supports the intuition, that in an environment with price-sensitive customers, the profit-optimal price tends to be lower. Even though the microeconomic profit-optimization is based on a monopolistic situation, it is essential to have this basic relationship in mind when thinking about price optimizations.

2.1.4 Terminology and Related Concepts

The terms price sensitivity and price elasticity are often used synonymously. More precisely, the concept of price sensitivity is broader than that of price elasticity. It includes customers' reaction to the price level as well as the reaction to price changes (Goldsmith et al. 2005, p. 501). Price sensitivity includes measures such as willingness to pay, price importance, price elasticity, switching behavior, price search behavior, and price consciousness (Lichtenstein/Ridgway/Netemeyer 1993; Zeithaml/Berry/ Parasuraman 1996; Monroe 2003).

Tellis (1988, p. 331) has a more narrow understanding of the concept and defines it as follows: „The term *price sensitivity* is a latent construct referring to the extent to which consumers vary their purchases of a product as its price changes". Overall, the price sensitivity of consumers is often captured via price elasticity calculations (Hamilton/East/Kalafatis 1997, p. 285; Hoch et al. 1995, p. 17). As price elasticities are "notoriously difficult to measure" (Ramirez/Goldsmith 2009, p. 199), self-reported measures of price sensitivity using a five point scale are utilized instead in some studies (Goldsmith/Flynn/Goldsmith 2003).

Price elasticities, however, are the most prevalent measure of price responsiveness in the marketing and economics literature (Mulhern/Williams/Leone 1998, p. 433). Especially the fact that price elasticity is a dimensionless indicator, and thus insensitive to the unit of measurement, makes it a suitable indicator to compare sensitivities across markets, products and model formulations (Bijmolt/Van Heerde/Pieters 2005, p. 141; Tellis 1988, p. 332). This research follows the understanding of price sensitivity being measured by price elasticity.

Because of the negative sign of the price elasticity, the terminology can be confusing. A "greater" price sensitivity means a more negative price elasticity and "less" price sensitivity means a less negative price elasticity (Tellis 1988, p. 332). Some research papers use the terminology "stronger" and "weaker" elasticities to avoid confusion (Ailawadi/Gedenk/Neslin 1999). If the terms "higher" and "lower" are used in this research, it is referred to the absolute value of the elasticity (cf. Danaher/Brodie 2000, p. 918 for this convention).

On the one hand price elasticity is part of microeconomic theory. On the other hand a broader understanding is developing in the academic field at the moment and price elasticity, being a measure of the reaction to price changes, is also part of the behavioral pricing field. This research stream analyzes how customers search for price information, how they process and evaluate it and how they use this information in their choice behavior (Homburg/Koschate 2005a; Homburg/Koschate 2005b). In general, behavioral pricing comprises price-related concepts like for example willingness to pay, price tolerance, price knowledge and reactions to price changes (Koschate 2002, p. 19). It is thus a perspective that complements classical price theory.

Overall, in order to understand empirical price elasticity data beyond their mathematical value, behavioral pricing concepts, such as price thresholds (Monroe 2003, p. 145-149) have to be taken into account. If a price change crosses a certain price threshold, the price elasticity will be greater than the price elasticity in the case where no threshold is crossed. To provide an example, a major brand of sparkling wine increased its price from 4.99 EUR to 5.49 EUR, crossing the psychologically important threshold of 5.00 Euro, which lead to an unusually high price elasticity of -3.63 compared to other price elasticities after similar price increases when no price thresholds were crossed (Simon/Fassnacht 2009, p. 163 f.). It is, however, not possible to determine how much of this effect can be attributed to crossing the price threshold and how much to the price increase.

Behavioral pricing concepts are usually hard to capture in econometric models; it is nevertheless essential not to examine price elasticities in isolation but to also bear in mind related concepts such as price thresholds (Monroe 2003, p. 145-149), customer satisfaction (Koschate 2002), price fairness (Xia/Monroe/Cox 2004; Herrmann et al. 2007), price tolerance (Herrmann et al. 2004; Wricke 2000, p. 6-8; Wricke/Herrmann/Huber 2000), reference price, prospect theory, loss aversion (Koschate 2002, p. 52-55; Madzumdar/Raj/Sinha 2005; Kucher 1985, p. 108-118; Bell/Lattin 2000; Han/Gupta/ Lehmann 2001) and price interest (Diller 2008, p. 101-120; Stamer/Liebermann 2004; Moosmayer/Wendlandt/Patz 2009). It might not be feasible to attribute how much of the price elasticity magnitude is caused by e.g. the price threshold effect, but managers and researchers must be aware of these related, behavioral concepts that are hard to grasp, in order to have a better understanding of the pricing environment and therefore the price elasticity per se. In the analysis at hand, price elasticities that are substantially influenced by behavioral concepts, such as price thresholds, will be pointed out.

2.2 Prior Research

Due to the importance of price elasticities within price management, researchers have reported a large number of econometric studies on price elasticities over the last decades. Even though comparing price elasticities can be problematic in some cases, e.g. due to the inconsistent calculation methodology and data sources (as described in chapter 2.1.2), a meta-analysis provides valuable insights.

The term meta-analysis was first introduced by Glass (1976) and refers to the "analysis of analyses"; particularly it refers to "the statistical analysis of a large collection of analysis results from individual studies for the purpose of integrating the findings" (Glass 1976, p. 3). The term meta-analysis encompasses various methods and techniques of quantitative research synthesis also developed by Smith/Glass (1977), Rosenthal/Rubin (1978) and Schmidt/Hunter (1977). A meta-analysis "is not a technique, rather it is a perspective that uses many techniques of measurement and statistical analysis" (Glass/McGaw/Smith 1981, p. 21). Therefore, a meta-analysis can be understood as a form of research in which research reports are surveyed. A coding form is developed, a sample of research reports is selected and the appropriate information about its characteristics and findings is coded. The resulting data are then analyzed using statistical techniques to examine and describe the pattern of findings (Lipsey/Wilson 2001, p. 1 f.). The academic use of meta-analyses has dramatically grown in the last decades (Hunter/Schmidt 2004, p. 24 f.).

Tellis (1988) provides an influential meta-analysis on price elasticities. The analysis covers 367 price elasticities published between 1961 and 1985. Bijmolt/Van Heerde/Pieters (2005) update Tellis' (1988) meta-analysis. The data is extended to 1851 price elasticities covering the time frame from 1961 to 2004. The selection of price elasticities is based on the following criteria:

- Elasticities of brand and stock-keeping-unit (SKU), not categories
- Elasticities of single brand or SKU, not averages across items
- Elasticities based on actual purchase or sales data, not data from experiments, surveys or expert judgment

- Elasticities derived from business-to-consumer market, not business-to-business market

The study of Bijmolt/Van Heerde/Pieters (2005) provides a systematic overview and summary of past research results and serves with Tellis (1988) as the main basis for the following chapters on the magnitude of price elasticity and the determinants of price elasticity.

2.2.1 Magnitude of Price Elasticity

The frequency distribution of the observed price elasticities of Bijmolt/Van Heerde/Pieters (2005) is presented in figure 2-2. The overall mean of the 1851 price elasticities is -2.62 with a median of -2.22 and a standard deviation of 2.21. 50% of the values lie between -1 and -3 and 81% between 0 and -4. The elasticities range between 4.00 and -18.90, no exact values on individual price elasticities are provided in the meta-analysis, only a frequency distribution across this range is displayed (Bijmolt/Van Heerde/Pieters 2005, p. 145). 2.2% of the elasticities are positive.

Figure 2-2: Frequency Distribution of Price Elasticities – 1961-2004

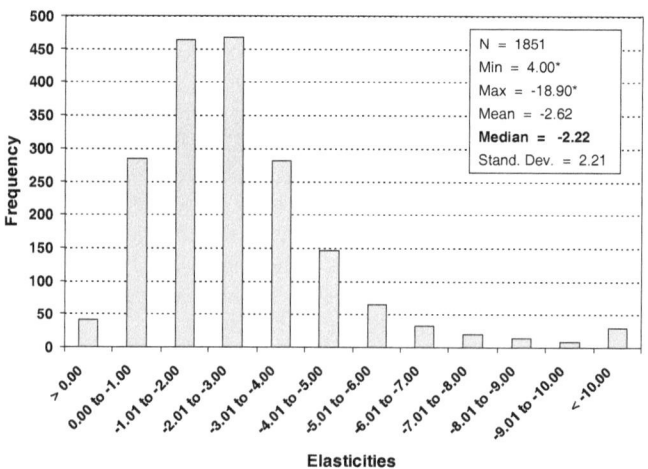

Source: adapted from Bijmolt/Van Heerde/Pieters (2005, p. 145)
*the price elasticities range from 4.00 to -18.90, no exact individual values are provided

Tellis (1988, p. 337) reports, based on 367 price elasticities, a lower average price elasticity, the mean is -1.76 and the standard deviation is 1.74. The mode, the value that occurs most frequently in the data set, is -1.50. The price elasticities concentrate around the mean and the mode. The median as well as the minimum and maximum value are not reported.

Tellis (1988, p. 334) is more detailed in the categorization of price elasticities as he distinguishes six product categories: detergent, durables, food, toiletries, pharmaceutical and others. Whereas Bijmolt/Van Heerde/Pieters (2005, p. 147) distinguish only three categories: low stockpiling groceries, high stockpiling groceries and durables. Thus more recent insights on differences in price elasticities for various product categories are rather limited. To the author's knowledge, a systematic academic overview on different product categories is nonexistent. A recent study provides mean price elasticities for 19 grocery products (Gordon/Goldfarb/Li 2013), as Hoch et al. (1995) provide mean elasticities for 18 grocery product categories, the elasticities in these two studies are, however, aggregated averages across categories and not calculated for individual products. In addition, Gordon/Goldfarb/Li (2013, p. 11) show that price elasticities are not constant and vary over time as they measured price elasticities over 6 years on a quarterly basis.

A broader overview of various product categories beyond fast moving consumer goods and across industries is still lacking in academic research. More managerial based data on price elasticities in various product categories can be found in an overview of Simon/Fassnacht (2009, p. 107).

This research aims to enhance the knowledge on the magnitude of price elasticity by further analyzing academic data for individual price elasticities, not averages across categories, and build an academic database that enables pooling individual elasticities for the categories previously analyzed and comparing price elasticities across studies as well as products and brands. In addition, it aims to build a database on price elasticities based on consulting project data to gain knowledge on a broader spectrum of products and industries.

2.2.2 Determinants of Price Elasticity

Again, the author refers to Bijmolt/Van Heerde/Pieters (2005) who comprise Tellis (1988) and represent the state of the art in research covering determinants on price elasticities. The analysis distinguishes two groups of determinants: research methodology (table 2-1) and market characteristics (table 2-2).

Table 2-1: Research Methodology: Effects on Magnitude of Price Elasticity

Determinant	Levels as defined by Bijmolt/Van Heerde/ Pieters (2005)	Bijmolt/Van Heerde/ Pieters (2005)	Tellis (1988)
Data source	Firm (ex-factory) vs. store panel vs. household panel	n.s.	n.s.
Temporal aggregation	Weekly/biweekly vs. monthly to yearly	n.s.	Weekly/biweekly > monthly to yearly
Item definition	SKU vs. brand	SKU > brand	n.a.
Criterion variable	Relative (market share, choice) vs. absolute sales	Absolute > relative	Absolute > relative
Functional form	Multiplicative/ exponential vs. attraction vs. additive	n.s.	n.s.
Definition of price	Actual price vs. promotional price	Short term: promotional > actual; Long term: actual > promotional	n.a.
Price endogeneity	Not acccounted vs. accounted for	Accounted > not accounted for	n.a.
Quality effect	Omitted vs. included	n.s.	Included > omitted
Distribution effect	Omitted vs. included	n.s.	Omitted > included
Advertising effect	Omitted vs. included	Omitted > included	n.s.
Sales promotion effect	Omitted vs. included	Omitted > included	n.s.
Estimation method	OLS vs. GLS….etc.*	n.s.	GLS < other methods*
Heterogeneity in price sensitivity	Not accounted vs. accounted for	n.s.	n.a.

n.s.: not significant, n.a.: not applicable
*methods tested include: OLS = ordinary least squares, GLS = generalized least squares, WLS = weighted least squares, SUR = seemingly unrelated regression, 2SLS = two-stage least squares, MLE = maximum likelihood estimation
comparisons refer to absolute magnitude of price elasticities
Source: adapted from Bijmolt/Van Heerde/Pieters (2005), p. 153

Chapter 2: Foundation of Research   21

The research methodology group comprises 13 methodology-oriented variables tested by Bijmolt/Van Heerde/Pieters (2005). Nine of these variables are also analyzed by Tellis (1988). Two thirds of these determinants lead to conflicting results, only one third leads to consistent results. This illustrates the limited ability to generalize results on price elasticities.

The three determinants with congruent results are the data source, the functional form and the criterion variable. Price and volume data can be obtained from different sources, specifically firm/ex-factory data, store panels or household panels. Both studies find no significant effect of the type of data source on the magnitude of price elasticity.

The price response function can have various forms. The meta-analyses distinguish between three categories of models to obtain price elasticities: multiplicative or exponential models, attraction models and additive models. Both studies find no significant effect of the functional form on the magnitude of price elasticity.

Findings on the criterion variable are also congruent. Criterion variable is the name Bijmolt/Van Heerde/Pieters (2005) chose for the type of sales measure: relative (market share and choice) vs. absolute sales. In both studies absolute sales elasticities are higher than relative (market share and choice probabilities) elasticities. Absolute and relative price elasticities are identical if price changes cause only brand switching but do not cause differences in purchase incidence and purchase quantity, which would lead to primary demand effects. Primary demand effects are not captured by relative price elasticities which explains them to be smaller in magnitude compared to absolute sales price elasticities. However, the congruent finding as displayed in table 2-1 adapted from Bijmolt/Van Heerde/Pieters (2005) is only true for the early stages of the product life cycle. Tellis (1988, p. 340) finds that price elasticities based on sales data are larger in magnitude in the early stages of the product life cycle. This is in line with Simon's (1979) argumentation that as sales are lower in the early stages of the product life cycle, the primary demand effects (capturing purchase incidence and purchase quantity) are stronger and can be better captured by sales as the criterion variable. Secondary demand effects capture brand choice and brand switching. The main effect of using absolute sales in the late stages has a negative effect on the magnitude of price elasticity (a positive effect on the actual value of price elasticity). Thus using absolute sales in the late stages of the product life cycle leads to less negative price elasticities (Tellis 1988, p. 338 f.). Bijmolt/Van Heerde/Pieters (2005, p. 146) also test for an interaction of the product life cycle and the criterion variable (absolute vs. relative sales) and do not find a significant interaction effect between these variables.

As a general note, it might be confusing to the reader that when Tellis (1988, p. 339) says "positive effects" to the actual price elasticity, while Bijmolt/Van Heerde/Pieters (2005, p. 153) refer to the magnitude of price elasticity. In the first case a positive effect lowers the absolute price elasticity (the negative value becomes less negative) and in the second case a positive effect increases the absolute price elasticity. This illustrates that the terminology with regard to price elasticities is not consistent in the literature. Another example that illustrates the inconsistency can be taken from Gordon/Goldfarb/Li (2013). In their study, the authors sometimes display price elasticities as

$$-\frac{\partial q(p)}{\partial p}\frac{p}{q},$$

so they generally appear as positive numbers rather than negative numbers (e.g. table 3, Gordon/Goldfarb/Li 2013, p. 11). The authors attempt to facilitate interpretation, especially when correlations with determinants are calculated (table 5, Gordon/Goldfarb/Li 2013, p. 16) based on

$$-\frac{\partial q(p)}{\partial p}\frac{p}{q},$$

but the authors are not consistent within their study since in the same table 5 (p. 16), the price elasticities are displayed as

$$\frac{\partial q(p)}{\partial p}\frac{p}{q}$$

and thus negative numbers, while the correlations shown in the table are based on calculations with

$$-\frac{\partial q(p)}{\partial p}\frac{p}{q}.$$

A positive correlation means than that the price elasticity rises with higher levels of the determinant, but again Goldfarb/Gordon/Li (2013) mean the absolute value when they write about rising price elasticity. The term price sensitivity is also used as a substitute throughout the text. However, if negative price elasticity values are used the opposite sign will appear and the correlation or effect will not be positive but negative and vice versa. It is understood that this can be very confusing. This example illustrates the importance of being aware of what data, notations and calculations were used when determinants are analyzed, so the data can then be interpreted accurately.

As mentioned before, there are conflicting results for the majority of the determinants. The temporal aggregation of the price and sales data, weekly or biweekly vs. monthly to yearly, could have an effect on price elasticities since aggregation over time leads to the loss of information on price and sales fluctuations. The level of aggregation has no significant effect in the study of Bijmolt/Van Heerde/Pieters (2005) but Tellis (1988) could demonstrate a negative effect for higher levels of aggregation on price elasticities. Price elasticities calculated on higher aggregation levels do not capture some short-term price changes and are therefore less elastic due to less information on the temporal component of price elasticities, i.e. short-term price changes. The omission or inclusion of the quality, distribution, advertising and sales promotion effect did not lead to congruent results, neither did the type of estimation method.

The remaining market characteristic variables are exclusively studied by Bijmolt/Van Heerde/Pieters (2005). The item definition (SKU vs. brand) has a significant effect and higher price elasticities are measured for SKUs in comparison to brands, this can be explained by the fact that intra-brand switching cannot be observed at the brand level, only the SKU level; therefore the price elasticities at the SKU level tend to be higher. Analyzing the impact of the price definition (actual vs. regular vs. promotional price) leads to the observation that in the short term the price elasticity is higher for the promotional than for the actual price, whereas in the long-term this relationship is reversed. The regular price could not be analyzed due to the limited number of observations. Accounting for endogeneity of prices leads to larger price elasticities, while no significant effect can be found for accounting for heterogeneity.

The second group of determinants represents market characteristics (table 2-2); these determinants include brand and category characteristics as well as economic conditions.

Table 2-2: Market Characteristics: Effects on Magnitude of Price Elasticity

Determinant	Levels as defined by Bijmolt/Van Heerde/ Pieters (2005)	Bijmolt/Van Heerde/ Pieters (2005)	Tellis (1988)
Brand ownership	Manufacturer brand vs. private label	n.s.	n.a.
Product category	Groceries low stockpiling, groceries high stockpiling, durables	durables > groceries	durables > food > pharmaceuticals
Stage of product life cycle	Introduction/growth vs. mature/decline	Introduction/growth > mature/decline	Mature/decline > introduction/growth
Country	U.S./Canada vs. Europe vs. Australia/New Zealand/Japan	n.s.	Australia/New Zealand > U.S. > Europe
Household disposable income	Linear effect	n.s.	n.a.
Inflation	Linear effect	Positive effect	n.a.
Year of data collection	Linear effect	Absolute sales elasticity: positive effect, relative elasticity: n.s.	n.a.

n.s.: not significant
n.a.: not applicable
effects and comparisons refer to absolute magnitude of price elasticities
Source: adapted from Bijmolt/Van Heerde/Pieters (2005), p. 153

Seven market characteristics are analyzed by Bijmolt/Van Heerde/Pieters (2005). The majority of these was not included in Tellis' (1988) analysis. Three are tested in both studies with conflicting results in two cases and some consistency in one case, again demonstrating the difficulty of generalizing results.

The consistency, which is however limited, is shown for the product category. Tellis (1988) distinguishes more categories (detergent, durable goods, food, toiletries, pharmaceuticals and others) than Bijmolt/Van Heerde/Pieters (2005) who differentiate groceries with low and high stockpiling propensities and durables. The overview in table 2-2 states that price elasticities for durables are larger than those for groceries (Bijmolt/Van Heerde/Pieters 2005) and that price elasticities for durables are larger than those for food which in turn are larger than those for pharmaceuticals (Tellis 1988). The statement of Bijmolt/Van Heerde/Pieters (2005, p. 152) "we reconfirm that consumers are more price elastic for durables than for other products" is misleading since for example the detergent category (which is comprised by Bijmolt/Van Heerde/Pieters under groceries) has a larger price elasticity than the durable category as shown by Tellis (1988, p. 334, 338).

No generalization is possible regarding the stage of product life cycle. Both studies distinguish early stages of introduction and growth and later stages of maturity and decline. Bijmolt/Van Heerde/Pieters (2005) find that price elasticities are larger in the early stages of the product life cycle than in the later stages. A higher degree of product differentiation and more loyal customers can cause lower price elasticities in later stages (Simon 1979). The tables lists the results for Tellis (1988) as reverse, i.e. that price elasticities are larger in the maturity and decline phases than in the introduction and growth phases. This can be explained by the fact that the broader customer base in the later stages has better information on availability, prices and promotions and thus are more price sensitive than the segment of highly involved early adopters and innovators (Ghosh/Neslin/Shoemaker 1983; Parker/Neelamegham 1997; Tellis/Fornell 1988). However, looking at the interaction effect between product life cycle and the sales measure (absolute vs. relative sales) Tellis (1988, p. 340) found that price elasticities are larger in magnitude in the early stages. This can be explained with Simon's (1979) argumentation that sales are generally lower in the introduction and growth stages; primary demand effects are stronger and can be better captured with absolute sales as the measure. Bijmolt/Van Heerde/Pieters (2005) do not find a significant interaction effect between these two variables.

The possibility to generalize is also limited for the national setting. Tellis (1988) finds the lowest price elasticities for Europe and the highest for Australia and New Zealand. Elasticities for the United States lie in the middle, and Japan was not included in the data set. The differences could not be explained by Tellis (1988, p. 339) and he suggests that the differences warrant further studies. The analysis of Bijmolt/Van Heerde/Pieters (2005) does not reveal significant differences in price elasticities. All data is obtained from highly developed areas which leads to a higher degree of homogeneity compared to a data set that includes less developed areas.

The remaining determinants were exclusively studied by Bijmolt/Van Heerde/Pieters (2005). The brand ownership with the levels manufacturer brand and private label shows no significant effect on price elasticity. Also, the household disposable income shows no significant effect. It could, however, be demonstrated that inflation has a positive effect on price elasticity, which means that in times of inflation customers tend to be more price sensitive. The overall time trend measured by the year of data collection has no significant effect on price elasticity (Bijmolt/Van Heerde/Pieters 2005, p. 146 f.), the table shows a significant interaction effect between the year of data collection and the criterion variable relative vs. absolute sales. There is a magnitude increasing effect on the absolute sales elasticity while there is no significant effect on the relative sales elasticity. For a more comprehensive discussion of both the research methodology and market characteristics determinants, see Bijmolt/Van Heerde/Pieters (2005).

3 Data

The data for this research is derived from two main sources, academic publications (chapter 3.1) and consulting projects (3.2). This enables the author to go deeper into the currently available data set published in academic journals and use this data in new ways such as an analysis of price elasticities by product category across studies and an analysis of brands within the product category.

Adding the consulting projects as a data source expands the available data primarily in two ways. It expands the range of product categories beyond fast moving consumer goods and allows the exploration and analysis of a currently underutilized data source in the academic research. Another main advantage is the fact that the data stems from real consulting projects indicating a high degree of practical relevance, which is always aimed for in further research. Having a more consistent data source and calculation methodology compared to the academic studies, the consulting project data will make comparisons across product categories more accurate.

3.1 Academic Publications

Academic publications are the first data source. In order to gain a thorough understanding of the currently available academic research data, a comprehensive review was conducted.

3.1.1 Data Collection

The data collection in the academic field was initiated with the publications included in the meta-analysis of Bijmolt/Van Heerde/Pieters (2005, pp. 143-144), as these authors had already identified studies that report price elasticities through an elaborate search strategy. As mentioned before, an observation of price elasticity was included by Bijmolt/Van Heerde/Pieters (2005, p. 142) if the following four criteria were fulfilled, in line with Tellis (1988):

- Price elasticities of brand and stock-keeping-unit (SKU), not categories

- Price elasticities of single brand or SKU, not averages across items

- Price elasticities based on actual purchase or sales data, not experimental or judgmental data

- Price elasticities derived from business-to-consumer market, not business-to-business market

This data set was narrowed down to journal articles published within the last 25 years (2006 to 1981) so as to work with more recent data compared to data starting in 1961 for Bijmolt/Van Heerde/Pieters' (2005) meta-analysis data. The publication date was used as a benchmark for the year of data collection. The publication has an average time lag of 8 years (Bijmolt/Van Heerde/Pieters 2005, p. 145). Outliers were identified and excluded if the time lag was too large, e.g. a study from the 1990s with data from the 1970s on the West German market (Wagner/Taudes 1991). Another reasoning for this cut is the fact that scanning was introduced in the 1970s and diffused quickly by 1980 (Bijmolt/Van Heerde/Pieters 2005, p. 151).

"Unlike some research areas in marketing (i.e. logistics, quality management, and to some degree services) that have been usurped by other disciplines (i.e. operations or management science), marketing appears to be on solid ground with respect to the pricing domain" (Leone et al. 2012, p. 1023). Therefore, the relevant data is substantially covered by concentrating on marketing journals.

To ensure a high level of quality and relevance, the analysis only includes studies from the top 25 marketing journals according to any of the following four rankings covering a broad range of evaluation methods. Bauerly/Johnson (2005) conducted a syllabi analysis from seminars in doctoral marketing programs of AACSB-International-accredited schools to evaluate journal quality. Baumgartner/Pieters (2003) analyzed citation exchanges among marketing journals to determine the structural influence of the journals. Hult/Neese/Bashaw (1997) asked marketing faculty of AACSB accredited and non-accredited schools, that are partly doctorate granting and partly non-granting, to rank order their top 10 most important journals. Theoharakis/Hirst (2002) interviewed the marketing faculty of leading business schools and assessed the journal familiarity, the average rank position of the journal, the percentage of respondents who classify a journal as top tier, and the percentage of respondents who regularly read the journal. In all four rankings, the Journal of Marketing ranks highest and the Journal of Marketing Research and the Journal of Consumer Research rank second or third. A detailed overview of the top marketing journals according to these rankings can be found in the appendix (cf. Appendix A: Top 25 Marketing Journals).

The combination of objective measures such as frequency of citations by the research community and subjective measures such as expert opinion surveys ensures that a broad array of evaluation criteria are covered and thus the influential journals are identified. If a journal does not belong to the top 25 marketing outlets in any of these four studies on journal quality, the price elasticity articles published in these journals were excluded. This procedure led to the elimination of 20 articles; the excluded journals were for example the American Journal of Agricultural Economics, the Agricultural and Resource Economics Review, the Agribusiness, the Journal of the American Statistical Association, and the International Journal of Forecasting. In the last step, the data was broadened by studies published between 2004 and 2006 fulfilling the described selection criteria.

3.1.2 Overview of Selected Studies

The data collection led to a revised and updated database comprising 46 academic articles and 863 price elasticities. The following table 3-1 provides an overview of the selected studies, including research focus, analyzed products and determinants, number of price elasticity observations as well as key findings with regard to price elasticities. Such an overview was not to be found in the academic research stream up until now.

Table 3-1: Selected Academic Studies on Price Elasticities

Authors(s) (Year)	Research Focus	Product(s)	Determinants Analyzed	# of PE	Key Findings Regarding Price Elasticities
Ailawadi/Gedenk/ Neslin (1999)	Model building to account for heterogeneity and purchase feedback	Yogurt Ketchup Saltine crackers	Different methods to account for heterogeneity and purchase feedback	30	o Accounting for heterogeneity and purchase feedback lowers PE o PE not sensitive to the method used (all models weaken PE roughly to the same degree)
Allenby (1989)	Identification, estimation and testing of demand structures	Toilet tissues	None analyzed with respect to PE	18	o No specific findings regarding PE
Allenby/Rossi (1991)	Quality perceptions and asymmetric switching between brands	Margarine	Quality	40	o PE varies widely depending on model used o High quality brands have very high elasticities compared to lower quality brands
Bemmaor (1984)	Threshold effects of advertising	Frequently purchased branded consumer good (unrevealed)	Advertising expenditure	20	o PE is larger for heavily advertised premium brands than for less advertised low-price brands
Besanko/Gupta/Jain (1998)	Accounting for price endogeneity	Yogurt Ketchup	Price endogeneity	12	o PE is downward biased when price endogeneity is not accounted for (lower PE)
Bolton (1989a)	Robustness of PE estimates	Frozen waffles Liquid bleach Toilet tissues Ketchup	Functional form of brand sales equations (linear, multiplicative, exponential)	31	o PE estimates from linear and multiplicative models are both overstated o Linear form overstates PE relative to multiplicative form

Chapter 3: Data

Authors(s) (Year)	Research Focus	Product(s)	Determinants Analyzed	# of PE	Key Findings Regarding Price Elasticities
Brodie/de Kluyer (1984)	Evaluation of different market share models	Chocolate biscuits	Functional form (linear, multiplicative, attraction)	18	o Linear and multiplicative models perform as well or even better than attraction model
Bucklin/Russell/ Srinivasan (1998)	Relationship between market share elasticities and brand switching probabilities	Laundry detergent	None analyzed with respect to PE (only cross-PE)	9	o Aggregate PE are proportional to one minus the aggregate repeat purchase probabilities
Carpenter et al. (1988)	Modeling asymmetric competition (focus on cross-PE)	Household product (unrevealed)	None analyzed with respect to PE (only cross-PE)	11	o Premium brands with tangible benefit to the customer have lower PE than economy brands
Chib/Seetharaman/ Strijnev (2004)	Modeling no-purchase and brand choice decisions simultaneously	Cola	None analyzed with respect to PE	8	o Category purchase accounts for 21.8% of brands' total PE o No consistent relation of PE and brand vs. private label
Chintagunta (1992)	Accounting for heterogeneity	Yogurt	Heterogeneity	8	o Not accounting for heterogeneity lowers PE
Chintagunta (1993)	Impact of marketing variables on category purchase, brand choice and purchase quantity decisions	Yogurt	Conditional on purchase and unconditional on purchase brand choice	12	o Unconditional brand choice PE are larger than conditional ones

Authors(s) (Year)	Research Focus	Product(s)	Determinants Analyzed	# of PE	Key Findings Regarding Price Elasticities
Chintagunta (2001)	Accounting for endogeneity and heterogeneity	Shampoo	Heterogeneity across consumers Endogeneity of price	39	o Not accounting for either endogeneity or heterogeneity lowers PE o Not accounting for endogeneity has bigger impact than not accounting for heterogeneity
Chintagunta/ Honore (1996)	Modeling marketing variables and heterogeneity	Saltine crackers	Different model specifications	16	o PE smaller from probit model than from random effects probit model o PE from logit model are smaller than from mixture of logits
Chintagunta/Jain/ Vilcassim (1991)	Modeling heterogeneity	Saltine crackers	None analyzed with respect to PE	4	o Market leader has lowest PE o Private label has also rather low PE
Christen et al. (1997)	Debiasing market level data	Detergent Peanut butter	Market vs. store level data	28	o Using market level data tends to overestimate PE o Suggested debiasing approach improves estimates
Cooper (1988)	Asymmetric competition	Coffee	None analyzed with respect to PE	12	o Market leader has highest PE, 80% of these sales are on price promotions o Everyday low price brands do not generate enough variation to achieve PEs of the more frequently promoted brands
Gönül/Srinivasan (1993)	Modeling heterogeneity	Disposable diapers	Multiple sources of heterogeneity	12	o Incorporating heterogeneity sources improves model fit o Incorporating heterogeneity sources increases PE by 50-80%

Chapter 3: Data

Authors(s) (Year)	Research Focus	Product(s)	Determinants Analyzed	# of PE	Key Findings Regarding Price Elasticities
Guadagni/Little (1983)	Calibration of logit model on scanner data	Coffee	None analyzed with respect to PE	16	o PE large share brands are lower than PE of small share brands o Promotional PE is lower than regular PE (due to calculation method)
Gupta et al. (1996)	Impact on brand choice: household vs. store scanner data	Detergent Peanut butter	Data source (household vs. store) Selection procedure	81	o PE of household data is significantly different from store data PE, small difference 5% and 7% and no clear direction of impact on magnitude o PE household selection is lower than PE purchase selection
Hildebrandt/ Klapper (2001)	Price competition between corporate brands	Body care (unrevealed)	None analyzed with respect to PE	9	o No specific findings regarding PE
Kadiyali/ Chintagunta/ Vilcassim (2000)	Power of channel members	Orange juice Tuna	None analyzed with respect to PE	6	o Private label has highest PE (orange juice)
Kalyanam (1996)	Pricing decisions under demand uncertainties	Coffee	Model specifications – doublelog vs. semilog	12	o PE doublelog < PE semilog model
Kamakura/Russell (1989)	Market segmentation	Unrevealed food item	None analyzed with respect to PE	4	o No specific findings regarding PE
Kim (1995)	Accounting for heterogeneity	Tuna	Aggregation level Heterogeneity	18	o Proposed model (HAL) improves prediction performance o PEs from HAL are close to the heterogeneous logit model applied to household level panel data ("truth")

Authors(s) (Year)	Research Focus	Product(s)	Determinants Analyzed	# of PE	Key Findings Regarding Price Elasticities
Kim/Allenby/Rossi (2002)	Modeling demand for variety	Yogurt	None analyzed with respect to PE	20	o PE inside good model < PE outside good model o High variation of PE across flavors of identical yogurt brand
Kim/Rossi (1994)	Heavy-user bias	Tuna	Purchase frequency Purchase volume Sample selection	10	o Consumers with high purchase frequency or high purchase volume are much more price sensitive than those with low frequency or low purchase volume o They also have a more defined preference for national brands o Households become more price sensitive via costly acquisition of price information
Kopalle/Mela/Marsh (1999)	Dynamic effects of discounting on sales	Liquid dishwashing Detergent	Promotion/discounting	6	o Temporary price reductions can increase price sensitivity o Promotions have positive contemporaneous effects on sales accompanied with negative future effects on baseline sales
Krishnamurti/Raj (1991)	Relationship between brand loyalty and price sensitivity	Coffee Unrevealed frequently purchased product	Customer loyalty	12	o Loyal customers are less price sensitive in the choice decision o But more price sensitive in the quantity decision o Overall, nonloyal customers are more price sensitive
Kumar/Divakar (1999)	Brand size competition	Potato chips Peanut butter	Different brand sizes of a brand	43	o PEs of brand sizes vary widely, even within the same brand

Authors(s) (Year)	Research Focus	Product(s)	Determinants Analyzed	# of PE	Key Findings Regarding Price Elasticities
Mantrala et al. (2006)	Optimal pricing strategies	Automotive hard parts	Quality level (good, better, best) Store characteristics	66	○ PE good < PE better < PE best quality ○ Store characteristics: geographic latitude increases PE, satellite stores have higher PE. No generalization is possible for the remaining 6 characteristics
Mehta/Rajiv/ Srinivasan (2003)	Formation of evoked set	Detergent	Evoked set of consumers	8	○ PE is underestimated if assumed that consumers get to know prices at zero cost ○ Consumer with high PE have larger consideration set than consumers with low PE
Montgomery (1997)	Micro-marketing pricing strategies	Orange juice	Customer demographics Competitive characteristics	11	○ Demographic and competitive variables explain 22% of the variance, competitive variables are less influential
Moon/Russell/ Duvvuri (2006)	Impact of reference price mechanism	Toilet tissues	Different reference price mechanisms of consumers	30	○ PE memory based reference price > PE no reference price > PE stimulus based reference price ○ Pattern holds true for price increase and decrease
Mulhern/Williams/ Leone (1998)	Influence of ethnic, income and brand determinant on PE	Liquor	Brand factors Consumer characteristics	14	○ Brand factors and customer characteristics explain 23% of the variance in PE

Authors(s) (Year)	Research Focus	Product(s)	Determinants Analyzed	# of PE	Key Findings Regarding Price Elasticities
Murthi/Srinivasan (1999)	Consumers' extent of information evaluation in brand choice	Ketchup	Consumers' extent of information evaluation	12	o Customers who engage in information evaluation have a higher PE o On more than 40% of purchase occasion price information is not evaluated
Reibstein/Gatignon (1984)	Optimal product line pricing	Eggs	None analyzed with respect to PE	14	o Variety of PE for different egg sizes o Private label has rather low PE but among other brands
Roy/Chintagunta/ Haldar (1996)	Incorporating purchase feedback, habit persistence and heterogeneity in dynamic brand choice models	Ketchup	None analyzed with respect to PE	12	o No specific findings regarding PE
Russell/Bolton (1988)	Relationship between market structure and elasticity structure	Bathroom tissues Bleach Ketchup Margarine	None analyzed with respect to PE	39	o Proposed model is a robust approach to predict PE structure o PE national brand > PE private label
Russell/Kamakura (1994)	Combining household (micro) and store (macro) level data	Detergent	None analyzed with respect to PE	30	o No specific findings regarding PE
Sivakumar (2001)	Category purchase in logit models	Ketchup	Category purchase (included vs. omitted)	8	o Ignoring the category purchase decision underestimates PE
Srinivasan/ Popkowski-Leszczyc/Bass (2000)	Impact of different type of price changes on MS and competitive reactions	Beer Margarine	Type of price change	16	o PE temporary price changes > PE evolving price changes

Chapter 3: Data

Authors(s) (Year)	Research Focus	Product(s)	Determinants Analyzed	# of PE	Key Findings Regarding Price Elasticities
Van Heerde/ Gupta/Wittink (2003)	Decomposition of price elasticity and unit sales	Yogurt Tuna Sugar	None analyzed with respect to PE	8	o Decomposition of PE leads to higher secondary demand effect estimations than decomposition of unit sales effects
Van Heerde/ Mela/Manchanda (2004)	Effects of innovation on market structure	Frozen pizza	Market entrance of innovator	12	o Introduction of innovative product increases PE of existing products
Villas-Boas/Winer (1999)	Accounting for endogeneity	Yogurt Ketchup	Endogeneity of marketing mix variables	12	o Not accounting for endogeneity may result in substantial bias of PE estimates o PE with endogeneity > PE without endogeneity
Villas-Boas/Zhao (2005)	Modeling supply and demand	Ketchup	Price endogeneity	6	o Accounting for endogeneity increases PE

The final time frame of the articles ranges from 1983 to 2006 compared to 1961 to 2004 in the meta-analysis of Bijmolt/Van Heerde/Pieters (2005), the data overlap with Tellis (1988) is limited since his studies range from 1961 to 1985. More precisely the studies start in 1962 not 1961 since no price elasticities are reported in 1961. Looking at the studies it becomes evident that most studies do not focus on price elasticities and their determinants. Some studies do not even explicitly report price elasticities and for the most part price elasticities are a by-product of the research.

16 out of the 46 studies do not analyze any determinants; this is more than a third of all studies. Almost all studies focus on research methodology and evaluate and compare different models, which increases the number of price elasticities tremendously and illustrates the focus on modelling in the research. Some studies even test ten or more models with three brands resulting in 30 and 39 price elasticity estimations for just three products (Ailawadi/Gedenk/Neslin 1999; Chintagunta 2001).

Thus the determinant group research methodology in Bijmolt/Van Heerde/Pieters' (2005) meta-analysis is covered in the original studies. Looking for example at accounting for price endogeneity or consumer heterogeneity, several studies explicitly study the impact of accounting versus not accounting for these factors (e.g., Ailawadi/Gedenk/Neslin 1999; Besanko/Gupta/Jain 1998; Chintagunta 1992; Chintagunta 2001, Chintagunta/Honore 1996; Chintagunta/Jain/Vilcassim 1991; Gönül/Srinivasan 1993; Kim 1995; Villas-Boas/Winer 1999; Villas-Boas/Zhao 2005).

Price elasticity itself is not the research focus in the majority of the studies. Only very few studies explicitly have price elasticities as their main research aspect. For, example, Allenby (1989) focuses on the identification, estimation and testing of demand structures. Some studies focus on brand switching, asymmetric competition and price competition between brands (Bucklin/Russell/Srinivasan 1998; Carpeter et al. 1988; Hildebrandt/Klapper 2001), while other studies focus on market segmentation (Kamakura/Russell 1989). Van Heerde/Gupta/Wittink (2003) study the decomposition of price elasticities but no determinants of price elasticities but rather a comparison of elasticity and unit sales decompositions; the magnitude of the price elasticities is only reported as a by-product of the research. Other studies look at data characteristic such as data source and aggregation level (Christen et al. 1997; Gupta et al. 1996; Russell/Kamakura 1994).

Even though market and product characteristics (time trend, manufacturer vs. private label, product category, stage of product life cycle, country, income, inflation rate) were coded as determinants in the meta-analysis of Bijmolt/Van Heerde/Pieters (2005), these determinants were not the research focus in any of the original studies. However, in rare cases factors, such as income, are studied in a broader context of influencing characteristics on the pricing strategies (Montgomery 1997; Mulhern/Williams/Leone 1998). The brand ownership (manufacturer brand vs. private label) is for example never explicitly studied as a determinant of price elasticity. Of course, additional insights and conclusions can be drawn when the data is reassessed; and more detailed overviews and analyses of the products, industries and determinants will be provided in chapter 4.1 and chapter 5.1.

In the appendix, additional details on the academic studies are provided. The information includes for example the item level assessed (SKU vs. brand level), the data source (which is primarily scanner based), the definition of the price and the kind of volume measurement (sales vs. market share). In addition, the methodology used to determine the price elasticities and the overall number of price elasticities per study are provided, it is pointed out if several models are tested for the same product cases, and finally the overall mean of the price elasticities per study is stated (cf. appendix table C: Selected Academic Studies on Price Elasticities – Further Information).

3.2 Consulting Projects

The second data source is data obtained from consulting projects. These projects were conducted by Simon-Kucher & Partners, a global consulting firm specialized in strategy and marketing. The company was founded by one of the leading pricing experts and a professor of marketing, Prof. Dr. Dr. h.c. mult. Hermann Simon, and is regarded as one of the world's leaders in giving advice on product pricing (Ewing 2004, p. 42; Toppin/Czerniaswka 2005, p. 97).

Projects are selected that comprise primary research on price response by means of expert interviews or customer surveys using either direct questioning or indirect methods such as conjoint analysis to generate a price response function. This data source provides several key advantages and is suitable to advance the knowledge on price elasticities. One advantage is the high degree of managerial relevance since the data is derived from actual consulting projects reflecting real business cases. Another advantage is the circumstance that all data comes from a single source, which ensures the use of precise academic methods.

An expansion of knowledge is given through the use of survey data since previous meta-analyses are based on real purchase data (Bijmolt/Van Heerde/Pieters 2005, p. 142; Tellis 1988, p. 332). A highly relevant approach to derive price response functions in the managerial world will be explored in this research. As hypothetical approaches currently seem to be underestimated in academic research (Miller et al. 2011, p. 182), the research at hand aims to gain knowledge from a currently underutilized data source.

In contrast to market data, survey data can be used to design pricing strategies for new products. Another advantage is the broad range of product categories that can be analyzed. Price elasticities in previous meta-analyses stem almost exclusively from fast moving consumer goods (98%, Bijmolt/Van Heerde/Pieters 2005, p. 147), which can be explained by the heavy usage of scanner data in academic research. The consulting project data offers access to a broad array of industries, since Simon-Kucher & Partners conducts pricing projects for a diverse set of clients covering different industries, such as automotive, pharmaceuticals, consumer goods, financial services, and telecommunication.

Another advantage is that there are fewer limitations regarding the way to calculate price elasticities based on a price response function. For example, a broader price range can be explored and a more uniform way of calculating price elasticities can be chosen in contrast to comparing price elasticities from diverse academic studies using very different approaches and different price spreads in determining price elasticities. Therefore, the consulting project data provides the opportunity to expand the current knowledge on price elasticities in several ways.

3.2.1 Data Collection

The goal for the data collection was to cover a broad range of products in various industries. First, Simon-Kucher & Partners' internal project database and interviews with industry experts within the company were used to identify suitable projects. Second, the following key words were used to screen the project database: adaptive conjoint analysis (ACA), choice based conjoint (CBC), conjoint, conjoint analysis, conjoint measurement, discrete choice model (DCM), discrete choice, new product pricing, new product pricing strategy, price, price increase, price elasticities, price elasticity, price level optimization, price optimization, price positioning, PRICESTRAT, pricing and reimbursement. Third, the subsequent consulting topics were analyzed: positioning, price elasticity modeling, price level optimization, price optimization, price structure optimization, pricing, pricing and reimbursement. In a last step, a chronological project by project analysis for the time period from July 2003 to July 2007 was conducted.

In order to be included in the database, projects had to report price elasticities either explicitly or indirectly through price response functions. The price elasticities had to stem from survey data, either expert judgment with Simon-Kucher's PRICESTRAT tool or customer interviews, e.g. with conjoint analysis or discrete choice modelling. PRICESTRAT is a decision support model developed by Simon-Kucher & Partners to structure expert interviews and to facilitate the assessment of consumers' reactions to price changes. Price elasticities derived from company sales or market share data as well as experimental data were excluded from the analysis to focus on the new data source of survey data compared to Bijmolt/Van Heerde/Pieters (2005, p. 142) who based their analysis exclusively on actual purchase and sales data and excluded elasticities based on experimental and judgmental data, such as purchase intentions and preferences. The survey data provides the opportunity to extend the knowledge on price elasticities based on a previously underutilized data source (Miller et al. 2011, p. 182). As marketing research demands a high degree of managerial relevance, this aim is ensured by analyzing data that was used in real business cases when actual companies asked for pricing advice regarding their products or services. In addition, the comparability of price elasticity data was enhanced by using a more consistent estimation method compared to previous research based on market or scanner data with limited und uncontrollable price variation. In addition, the elasticities had to represent the sensitivity of a single brand or SKU, therefore averages across items were excluded from the research.

Initially, 140 potential projects were identified. For each project, the project report and project documentation was reviewed to check if price elasticities are reported or if they can be derived from price response functions. The final data set consists of 62 projects matching the selection criteria.

Since there are various methods to calculate price elasticities, the calculation method used in this research is explained. In general, the arc elasticity of a price increase of 10% as well as the arc elasticity of a price decrease of 10% was calculated. The current price served as the price anchor in most cases. This methodology allows differentiating between the directions of the price change. The average price elasticity is the mean of the elasticities for the price increase of 10% and the price decrease of 10%. The original price points as studied in the customer surveys and expert interviews were used as the basis for the price elasticity calculations.

Regarding the delta of the price change, a spread of 10% was selected to have a large enough price change in the permanent, regular price to cause customers' reactions. Danaher/Brodie (2000, p. 924) model arc elasticity as a function of price change in both directions and show that as price changes become larger in magnitude, the price elasticities become approximately constant, this means that beyond a certain threshold, price elasticities do not vary with relative price change. The price change threshold identified varied between 3% and 10%, and averaged 5%. The phenomenon of a range of acceptable price differences around the brand's nonpromotional price is consistent with Kalwani/Yim's (1992) finding that price changes of 5% or less do not cause a significant change in customers' price perception. This is in line with Raman/Bass (1988) and Gurumurthy/Little (1989). Thus, these research results support the decision to use a 10% price change to calculate price elasticities. Overall, the original data from the primary research in the consulting projects is used with the corresponding price points assessed during customer or expert interviews. Other researchers like Sivakumar (2001) also use a price cut of 10% to analyze the impact on market share.

Previous studies examine mostly promotional and short-term price changes and therefore price cuts when calculating price elasticities (Anderson/Song 2004; Bijmolt/Van Heerde/Pieters 2005; Chandran/Morwitz 2006; Lu/Moorthy 2007), especially given the circumstance that most data is derived from fast moving consumer goods, price increases are rarely used to calculate price elasticities (Chintagunta/Honore 1996). In the research at hand, the focus will be on regular prices and long-term effects of price changes addressing the current lack of information in this area. For example, in the meta-analysis of Bijmolt/Van Heerde/Pieters (2005) only 1% of the price elasticity cases are based on regular prices. The price changes assessed in the consulting projects selected did not address temporary price changes such as promotions and short-term effects but focused on regular prices and long-term effects.

3.2.2 Overview of Selected Projects

The data collection led to 62 projects matching the selection criteria. The following table 3-2 provides an overview of the selected projects and states the project focus, the industry and product as well as the number of price elasticities. Due to confidentiality agreements with clients, the information is disguised and no further information about the projects can be reported in this overview.

The data set consists of 440 price elasticities. Five price elasticities were excluded since they are outside the interval of the mean price elasticity plus or minus five times the standard deviation, so the final data set consists of 435 price elasticities.

For these 435 individual cases, there are 415 elasticities for a price decrease of 10%, 409 elasticities for a price increase of 10% and 386 elasticities with data for both price decrease and increase. This can be explained by the fact that in some projects only price changes in one direction were analyzed. For example, in an industry with a long-time downward trend in prices like the telecommunications industry, the option of a price increase was not tested for all cases. Overall, there are 1210 price elasticity cases available for further analysis.

Chapter 3: Data

Table 3-2: Selected Consulting Projects on Price Elasticities

Year	Project Focus	Industry	Product(s)	# of PE
2001	Price optimization	Away from home food market	Food and beverages	44
2002	Price optimization	Logistics	Tunnel toll	1
2002	Price positioning	Automotive	Automobile	1
2002	Price positioning	Automotive	Automobile	1
2003	Exploring price potential	Automotive	Automobile	30
2003	Optimization of pricing process	Industrial goods	Crane part	1
2003	Optimization of pricing processes and price strategy	Industrial goods	Cranes and crane parts	3
2003	Price optimization	Industrial goods	Label printing machine	3
2003	Price positioning and market volume estimation	Industrial good	Hammer drill	1
2003	Price to lead	Automotive	Automotive tires	26
2003	Price optimization	Pharmaceuticals	Medications	3
2003	Power pricing	Industrial goods	Service contracts for elevators	1
2003	Price optimization	Industrial good	Self-adhesive product	1
2003	Pricing	Financial services	Financial services	4
2004	Repositioning	Industrial goods	Respirators	12
2004	Product and price optimization	Financial services	Insurance	4
2004	Pricing and reimbursement strategy	Pharmaceuticals	Medication	6

Year	Project Focus	Industry	Product(s)	# of PE
2004	Price optimization	Logistics	Airway cargo	2
2004	Value based power pricing	Industrial good	Die casting machine	1
2004	Pricing for key accounts	FMCG	Cosmetics	15
2004	Price optimization	Logistics	Parcel services	6
2004	Price optimization	Tourism	Ski rental	1
2004	Price optimization	Automotive	Automobile	1
2004	Price optimization	Automotive	Automobile	1
2004	Price optimization	Media	Newspaper advertising	8
2004	Price optimization	Automotive	Automobile	3
2004	Power pricing	Services	Service contracts for elevators	1
2004	Pricing opportunities	Pharmaceuticals	Medication	3
2004	Pricing and ensuring profitable growth	Technology	Technology	2
2005	Price optimization	Engineering	Stick electrodes and flux cored wires	4
2005	Price optimization	Engineering	Electrodes and welding equipment	9
2005	Pricing and reimbursement strategy	Pharmaceuticals	Medication	2
2005	Price optimization	Industrial goods	Film for printing/graphics	8
2005	Price optimization	Pharmaceuticals	Medication	1
2005	Price optimization	Logistics	Mail services	1
2005	Price optimization	Logistics	Parcel services	15
2005	Price optimization	Industrial goods	Copper crim fittings	2
2005	Price optimization	Industrial goods	Copper crim fittings	4

Chapter 3: Data　　　　　　　　　　　　　　　　　　　　　　　　　　　　　　　　　　　　45

Year	Project Focus	Industry	Product(s)	# of PE
2005	Price optimization	Automotive	Automotive spare parts	6
2005	Price optimization	Media	Newspaper subscription	2
2005	Future pricing strategy	Logistics	Train transportation	1
2005	Price optimization	FMCG	Food and beverages	4
2006	Price optimization	FMCG	Food and beverages	10
2006	Price optimization	FMCG	Home food packaging	17
2006	Price optimization	Logistics	Tunnel toll	4
2006	Price optimization	Automotive	Automobile	4
2006	Price optimization	Industrial goods	Rotor spinning machine	2
2006	Price optimization	Logistics	Mail and parcel services	6
2006	Price optimization	Industrial goods	Self-adhesive products	18
2006	Price optimization	Telecommunication	Telecommunication packages	3
2006	Optimizing pricing structure	Consumer durables	Glassware and tableware	28
2006	Price optimization	Consumer durables	Kitchen electronics	1
2006	Price optimization	FMCG	Groceries	14
2006	Price optimization	Technology	Technology	1
2007	Price optimization	Automotive	Automobile	1
2007	Pricing and reimbursement strategy	Pharmaceuticals	Medication	12
2007	Price test	Telecommunication	Telecommunication packages	25
2007	Pricing process and price strategy optimization	Medical technology	Medical device	12
2007	Price optimization	Pharmaceuticals	Dermal fillers	20

Year	Project Focus	Industry	Product(s)	# of PE
2007	Portfolio pricing	Consumer durables	Bed mattresses	5
2008	Price optimization	Technology	Technology	2
2008	Price optimization	Financial services	Financial services	5

Chapter 3: Data 47

The projects stem from 2001 to 2008 and cover a broad array of industries and products comprising, for example, automobiles, cranes, industrial machinery, electrodes, logistics and financial services, medications and cosmetics. A more detailed overview and analysis will be presented in chapter 4.2 and chapter 5.2.

Generally speaking the focus of most projects is price optimization which can be explained by the selection criteria. 61% of projects and 55% of the price elasticities stem from projects that explicitly state price optimization as their focus. Price positioning and repositioning are closely related consulting project themes. In the pharmaceutical and medical technology industry the projects are concerned with a pricing and reimbursement strategy. The optimal price identified ensures both the uptake by physicians as well as the reimbursement by insurance companies because without reimbursement most medications or medical devices will not gain substantial market share. Other projects look at a product portfolio and optimize the pricing structure of products in various product lines, e.g. lower-end and higher-end product lines. For example, one project optimized the pricing structure for glassware and tableware lines with a variety of designs, another project optimized the pricing structure of bed mattresses with the high-end model being priced 5 times the price of the entry model.

So in a broader sense, the main objectives of the selected projects are to find the optimal price for a product, under various secondary objectives such as ensuring profitable growth or optimizing the pricing process. Price elasticities play a central role in the price-optimization process; they indicate the level of price sensitivity of the consumers. In most cases a price response function is derived via survey data, which then serves as the basis for the price optimization. The price response function in turn provides the information needed to calculate the price elasticities.

4 Magnitude of Price Elasticity

In order to obtain a more comprehensive understanding of the magnitude of price elasticity both data sources, the academic publications (chapter 4.1) and the consulting projects (chapter 4.2), are utilized. The first step was to conduct a detailed study by study review in both areas to create a database on the magnitude of price elasticity. Detailed information on the price elasticities were recorded, particularly on the type of product and the brand, to enhance the current knowledge on the magnitude of price elasticities by being able to perform more detailed analyses, especially by product category.

4.1 Insights from Academic Publications

For each study the reported price elasticities were identified and recorded including further information on product category, brand and brand sizes. This enabled the author to show a general overview of the price elasticity data (chapter 4.1.1) as well as more specific overviews by product category (chapter 4.1.2).

4.1.1 Overview and Analysis of Price Elasticity Data

The academic data set consists of 863 price elasticities. The frequency distribution in figure 4-1 and the descriptive values are close to those of Bijmolt/Van Heerde/Pieters (2005). The mean price elasticity is -2.51 with a median of -2.21 and a standard deviation of 1.81. On average, the change in demand is two and a half times the change in price; given a 10% price reduction demand will increase by 25.1%. Half of the data (50.8%) lie between -1 and -3.

Chapter 4: Magnitude of Price Elasticity 49

Figure 4-1: Frequency Distribution of Price Elasticities – Academic Data Set

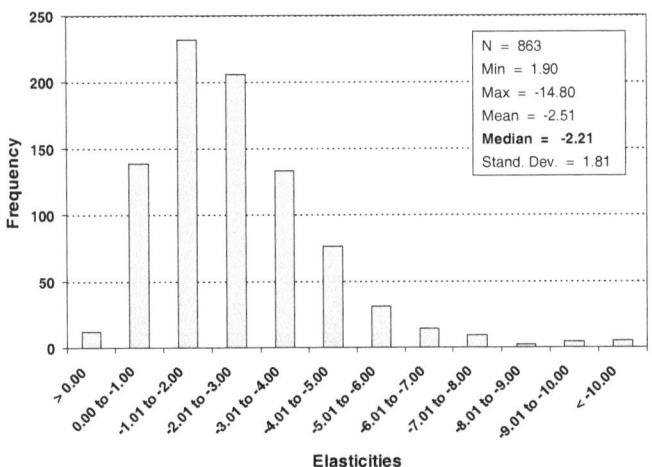

The Amoroso-Robinson relation (Simon/Fassnacht 2009, p. 207) indicates a potential to maximize profits for the 137 cases (15.9%) between 0 and smaller in magnitude than -1. There might also be a potential to raise prices for the positive price elasticity values (1.4%). The product category analysis will provide further information to identify these potentials.

In order to gain a better understanding of the price elasticities, the product categories are analyzed in detail. Bijmolt/Van Heerde/Pieters (2005) distinguish between groceries with low stockpiling propensities and groceries with high stockpiling propensities as well as durables. There are no further details on the actual products included in the study provided. Therefore, in the research at hand each of the 46 studies were assessed to gain detailed information on each of the 863 price elasticities regarding the product type, the brand and the brand size. This allowed creating an overview of the different products represented in the data set and comparing products and brands across studies.

Table 4-1 provides a more detailed overview on product categories compared to previous research. The products are almost exclusively fast moving consumer goods. An exception is the category of automotive hard parts (Mantrala et al. 2006) which was added in the process of extending the data set. It represents a special case of the do-it-yourself market in a predominantly business-to-business environment. Price elasticities on durables are not found in the meta-analysis of Bijmolt/Van Heerde/Pieters (2005) when the data is limited to the last 25 years.

About a third (32%) of the more recent academic data consists of price elasticities from the following three categories: laundry detergent, ketchup and bathroom tissues. Adding automotive hard parts, yogurt, margarine and peanut butter already covers 61% of all price elasticity cases. This clearly demonstrates how current research results are based on a rather limited amount of product categories. The remaining price elasticities stem from fast moving consumer goods, which Bijmolt/Van Heerde/Pieters comprise under the grocery category, e.g. products such as shampoo, orange juice and potato chips.

Table 4-1: Price Elasticities by Product Category in Academic Data Set

	Product Category	Total Number of PE	Product Category	Total Number of PE
	laundry detergent	127	orange juice	14
	ketchup	80	liquor	14
	bathroom tissues	66	eggs	14
61%	automotive hard parts	66	diapers	12
	yogurt	64	pizza	12
	margarine	62	bleach	11
	peanut butter	61	potato chips	11
	unrevealed consumer product	50	beer	9
	coffee	46	cola	8
	shampoo	39	waffles	8
	tuna	33	dishwashing detergent	6
	saltine crackers	30	sugar	2
	biscuits	18		
	Total = 863			

Table 4-2 shows a more detailed overview of the same product categories found in the academic data set, listing the number of elasticities sorted by product for each study separately. The total number of price elasticities for each product category is therefore the same as in table 4-1. However, it becomes apparent how many studies were conducted using this research subject. The Roman numbers (study I, study II, etc.) indicate the quantity of individual studies conducted for each product category. The number of studies listed in this table is higher than the number of academic publications since many authors study more than one product category in their research. Taking laundry detergent as an example, the table shows that there are five independent studies with an overall number of 127 price elasticities, the sample size ranges from 8 to 60 price elasticities per study. The number of price elasticity cases is listed in alphabetical order of the study authors' names being read from left to right in each row (in this example: Bucklin/Russell/Srinivasan: 9 price elasticities; Christen et al. 1997: 20 price elasticities; Gupta et al. 1996: 60 price elasticities; Mehta/Rajiv/Srinivasan 2003: 8 price elasticities; Russell/Kamakura 1994: 30 price elasticities). For the next product category (ketchup), there are nine different studies

Chapter 4: Magnitude of Price Elasticity

covering price elasticities and for bathroom tissues, there are four different studies covering price elasticities and so on. Overall, all studies included in table 4-2 are covered in the data overview table 3-1 (selected academic studies on price elasticities), where each academic study is listed in detail.

Table 4-2: Price Elasticities by Product Category and Study in Academic Data Set

Product category	Number of PE cases listed for each study within the product category separately									Total number of PE
	study I	study II	study III	study IV	study V	study VI	study VII	study VIII	study IX	
laundry detergent	9	20	60	8	30					127
ketchup	10	8	9	12	12	9	8	6	6	80
bathroom tissues	18	9	9	30						66
automotive hard parts	66									66
yogurt	10	4	8	12	20	4	6			64
margarine	40	15	7							62
peanut butter	8	21	32							61
unrevealed consumer product	20	11	9	4	6					50
coffee	12	16	12	6						46
shampoo	39									39
tuna	3	18	10	2						33
saltine crackers	10	16	12	6						30
biscuits	18									18
orange juice	3	11								14
liquor	14									14
eggs	14									14
diapers	12									12
pizza	12									12
bleach	5	6								11
potato chips	11									11
beer	9									9
cola	8									8
waffles	8									8
dishwashing detergent	6									6
sugar	2									2
Total										863

The highest number of studies is conducted for ketchup with nine different studies analyzing price elasticities for this product category leading to 80 price elasticities on this research subject. Laundry detergent has the largest number on price elasticity cases (127) that stem from a relatively small number of studies (5) compared to the ketchup category. It also becomes apparent that sometimes the overall number of cases for a product category is influenced by a single study examining a large number of cases, e.g. one study with 60 cases on laundry detergent (Gupta et al. 1996). In the case of automotive hard parts (66 cases) and shampoo (39 cases), the total number of price elasticities for these products stem from a single study within that product category.

4.1.2 Analysis of Selected Product Categories

In this part of the research, price elasticity distributions for specific product categories are presented. The product categories are shown in order of decreasing sample size, the cut-off in sample size is 30 price elasticity cases. This results in eleven product categories for a more detailed analysis, the remaining categories comprise 18 or less cases (cf. table 4-1). This chapter focuses on the descriptive values and the frequency distributions in the selected product categories and not on the determinants potentially causing these patterns.

Figure 4-2 illustrates the laundry detergent category which contains the largest amount of price elasticities – 127 cases. The distribution is strongly peaked with 60.6% of the observations being between -1 and -2. The mean is -1.93 with a median of -1.81 and a standard deviation of 0.78. This indicates that the relative volume change caused by a price change is about twice the relative price change. On average, a price decrease of 10% will lead to a volume increase of 19.3%.

Figure 4-2: Frequency Distribution of Price Elasticities – Laundry Detergent

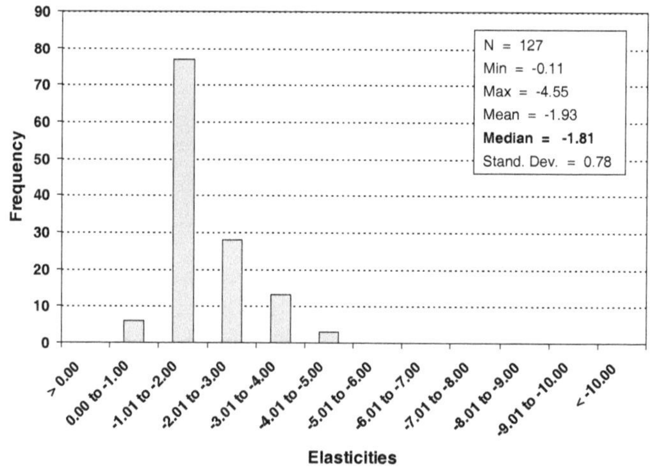

Out of the 46 research articles in this data set, nine studies (Ailawadi/Gedenk/Neslin 1999; Besanko/Gupta/Jain 1998; Bolton 1989a; Murthi/Srinivasan 1999; Roy/Chintagunta/Haldar 1996; Russell/Bolton 1988; Sivakumar 2001; Villas-Boas/Winer 1999; Villas-Boas/Zhao 2005) use ketchup as the research subject. This makes ketchup the most popular product used for price elasticity research (cf. table 4-2). Figure 4-3 shows the price elasticity distribution for the ketchup category. The distribution of the 80 price elasticities is broader compared to the laundry detergent category, that comprised data from five separate studies (Bucklin/Russell/Srinivasan 1998; Christen et al. 1997; Gupta et al. 1996; Mehta/Rajiv/Srinivasan 2003; Russell/Kamakura 1994). 31.3 % of the observations lie between -2 and -3 and 73.8% between -1 and -4. The mean is -3.03 with a median of -2.85 and a standard deviation of 1.42. The mean indicates an average volume change that is three times as high as the price change. However, the broader distribution pattern and the larger standard deviation indicate more variation in the price elasticities than in the previous product category laundry detergents. The strongest peak between -2 and -3 comprises 31.3% of the data, which is half the amount compared to the peak of 60.6% for laundry detergents between -1 and -2. These examples illustrate the various distribution patterns of price elasticities.

Figure 4-3: Frequency Distribution of Price Elasticities – Ketchup

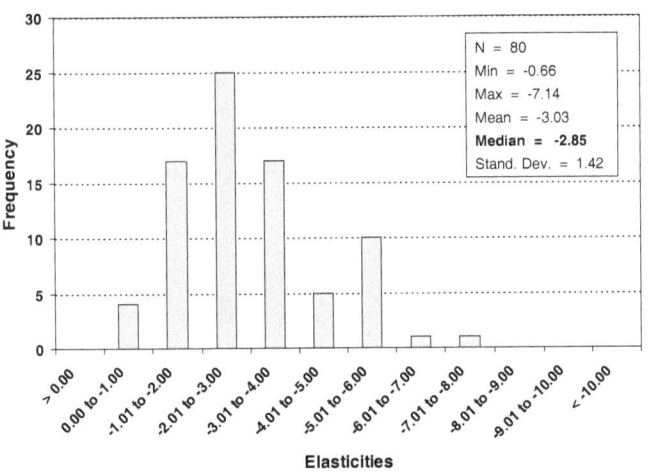

The broad amount of studies provides an opportunity to look more in detail at certain brand sizes to evaluate if it is possible to determine the price elasticity as a single representative value. Three studies (Besanko/Gupta/Jain 1998; Murthi/Srinivasan 1999; Roy/Chintagunta/Haldar 1996) estimate elasticities for the brand sizes Heinz 28oz., Heinz 32oz., and Hunts 32 oz. Table 4-3 shows the descriptive values for these brand sizes.

Table 4-3: Price Elasticities of Selected Ketchup Brand Sizes

Ketchup Brand Size	# of Price Elasticities	Min	Max	Mean	Median	Standard Deviation
Heinz 28oz	7	-0.66	-3.73	-1.88	-1.68	0.73
Heinz 32oz	7	-1.44	-3.52	-2.43	-2.64	1.04
Hunts 32oz	7	-2.13	-4.26	-3.20	-3.53	0.76

Source: own analysis based on data from Besanko/Gupta/Jain 1998; Murthi/Srinivasan 1999; Roy/Chintagunta/Haldar 1996

It becomes evident that there is not one representative price elasticity value for a product, not even for a specific brand or brand size of a product. This confirms Gabor's (1988, p. 18) statement that "...there is no such thing as *the* elasticity of demand for a good...". Instead, there is a broad range of price elasticities for each brand size. Even the brand size with the narrowest price elasticity range, i.e. Hunts 32oz., has a minimum price elasticity of -2.13 and a maximum that is twice that high (-4.26). The maximum price elasticity for Heinz 28oz. is five times as large as the minimum. This demonstrates that even for a very specific brand size of a product there can be a wide distribution of price elasticities.

The next product category in order of decreasing sample size (n = 66) is bathroom tissues. Compared to the previous products, bathroom tissues have a larger price elasticity with a mean of -4.09 (median -4.03, SD 1.60). The frequency distribution in figure 4-4 shows a concentration of 71.2% of the observations between -2 and -5. It could be that bathroom tissues are more price elastic since they are items with stockpiling ability bought frequently when on sale.

Chapter 4: Magnitude of Price Elasticity 55

Figure 4-4: Frequency Distribution of Price Elasticities – Bathroom Tissues

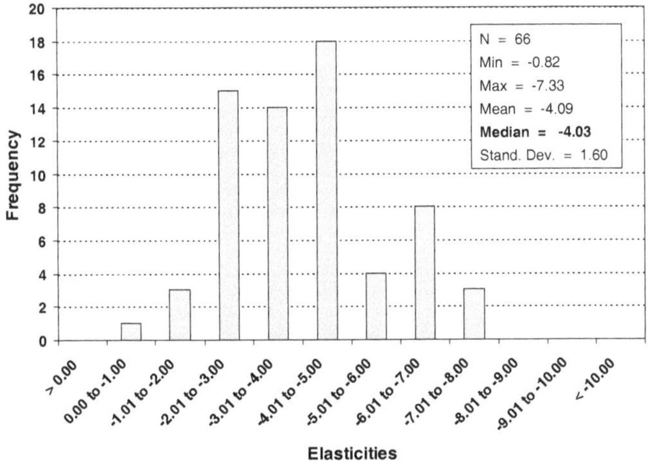

[Histogram showing frequency distribution of elasticities with statistics: N = 66, Min = -0.82, Max = -7.33, Mean = -4.09, Median = -4.03, Stand. Dev. = 1.60. X-axis shows elasticity bins from > 0.00 to < -10.00; Y-axis shows Frequency from 0 to 20.]

The elasticity distribution of automotive hard parts (figure 4-5) shows a different pattern, it is not bell-curved but concentrated towards 0. The distribution peak with a third of all elasticities lies between 0 and -1 and frequency of elasticities then declines for higher magnitudes of price elasticities. The 66 price elasticities range from -0.04 to -9.48. The mean is -2.22 (median = -1.71, SD = 1.83). One explanation for this pattern could be that for this product category the customer enters the market infrequently and buys only one particular hard part that is needed as a replacement. The product categories are composed of several subcategories where only one subcategory fits a particular make, model and year of a vehicle. The need for the replacement and the limited number of alternative choices could explain a lower price sensitivity of consumers. It also seems that there is a potential to raise prices for the inelastic automotive hard parts, i.e. the ones with a magnitude smaller than -1. The percentage change in quantity is less than the percentage in price. Thus, when the price is raised, the total revenue rises. The production of automotive hard parts is, however, not without costs and most of the time the overall goal of a company is to maximize not revenue but profit. A price increase always increases profits when the price elasticity magnitude is smaller than one, independently of the costs (Simon/Fassnacht 2009, p. 206).

Figure 4-5: Frequency Distribution of Price Elasticities – Automotive Hard Parts

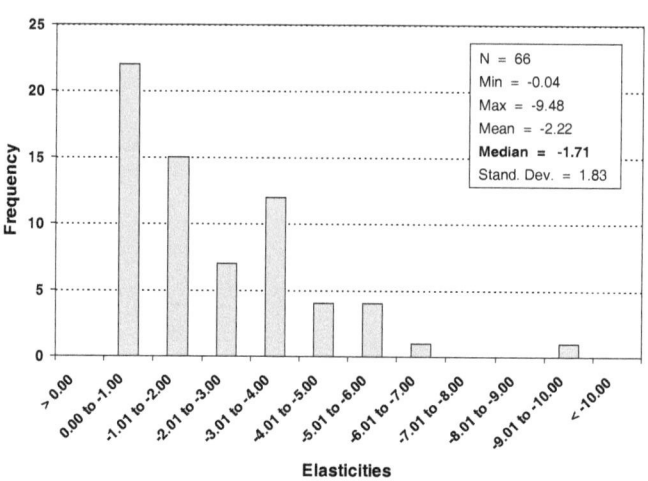

Another product category suitable for a deeper analysis is yogurt. The descriptive values and the frequency distribution are displayed in figure 4-6. There are 64 elasticity cases, the mean is -2.75 (median = -2.28, SD = 2.03). A third of all observations lie between 0 and -1 and two thirds between -1 and -3.

Chapter 4: Magnitude of Price Elasticity 57

Figure 4-6: Frequency Distribution of Price Elasticities – Yogurt

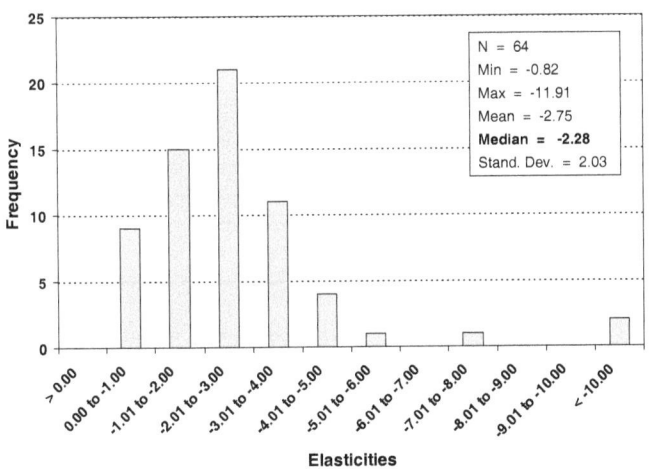

The data provides the opportunity to compare yogurt brands over six different studies (Besanko/Gupta/Jain 1998; Chintagunta 1992; Chintagunta 1993; Kim/Allenby/Rossi 2002; Van Heerde/Gupta/Wittink 2003; Villas-Boas/Winer 1999). Table 4-4 shows the descriptive values for the major national brands Dannon, Yoplait, Weight Watchers and Highland. Even though the mean price elasticity of Dannon, Yoplait and Highland are rather similar with -2.92, -2.78 and -2.65 respectively, the elasticity ranges are considerably different. The minimum price elasticity for Dannon is -0.82 and the maximum -10.50, whereas the range for Weight Watchers is much narrower (-0.90 to -4.10).

Chapter 4: Magnitude of Price Elasticity

Table 4-4: Price Elasticities of Selected Yogurt Brands

Yogurt Brand	# of Elasticities	Min	Max	Mean	Median	Standard Deviation
Dannon	29	-0.82	-10.50	-2.92	-2.42	1.85
Yoplait	9	-0.84	-11.91	-3.99	-2.91	3.61
Weight Watchers	7	-0.90	-4.10	-2.78	-2.95	1.19
Highland	5	-0.98	-4.87	-2.65	-2.09	1.52

Source: own analysis based on data from Besanko/Gupta/Jain 1998; Chintagunta 1992; Chintagunta 1993; Kim/Allenby/Rossi 2002; Van Heerde/Gupta/Wittink 2003; Villas-Boas/Winer 1999

In the next step the analysis looks at just one brand size, i.e. Dannon 8oz container, to narrow the research subject even further down from the yogurt category to a specific yogurt brand to finally a specific brand size. The price elasticities for this brand size are shown in table 4-5.

Table 4-5: Price Elasticities of Dannon Yogurt 8-ounce Size

Flavor	Market Share	Price Elasticity
Plain	12%	-1.88
Strawberry	33%	-2.21
Blueberry	18%	-2.26
Pina Colada	19%	-2.56
Mixed Berry	19%	-3.53

Source: adapted from Kim/Allenby/Rossi 2002

It becomes evident that even within one brand size analyzed in one study (Kim/Allenby/Rossi 2002) there is a high variation in price elasticities across flavors. The price elasticity of the mixed berry flavor (-3.53) is about twice as high as the elasticity of the plain flavor (-1.88). The market share does not explain the variation in price elasticity. Also, the price is identical for each flavor and therefore cannot explain the differences in price elasticity. Given that price elasticities vary significantly within one brand, it is unfeasible to generalize price elasticities for a product category.

There are 62 price elasticities for margarine (figure 4-7), the mean price elasticity is -3.26 (median = -3.29, SD = 1.67). Approximately three quarters (74.2%) of all observations lie between -2 and -5. This distribution pattern is similar to the pattern of bathroom tissues. However, there seems to be not much similarity in consumers' purchase behavior of margarine and bathroom tissues. Margarine has in contrast to bathroom tissues no stockpiling propensity; overall it is questionable if stockpiling propensity has an effect on price elasticities. Bijmolt/Van Heerde/Pieters (2005, p. 146) did not find an effect of stockpiling propensity on the elasticities in mature product categories. The mean of -3.26 is lower than that of bathroom tissues (-4.03).

Figure 4-7: Frequency Distribution of Price Elasticities – Margarine

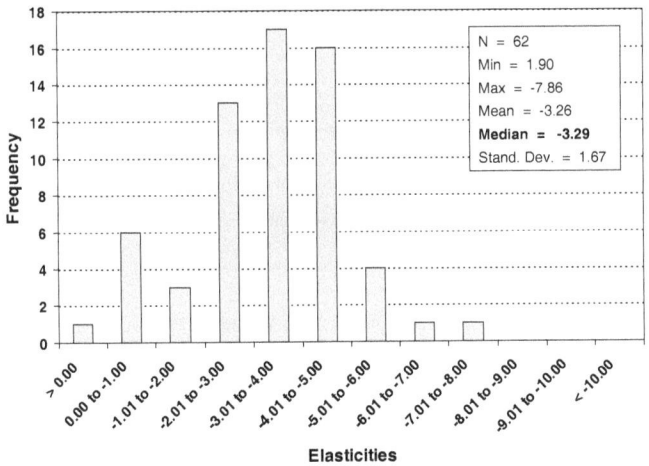

The elasticity distribution of peanut butter (n = 61) shows a very unique pattern (figure 4-8). The elasticities are strongly peaked between 0 and -1 as well as between -2 and -3 with 47.5% and 37.7% of the cases respectively. This rather unusual pattern can be explained by the fact that the data is derived from three studies (Christen at al. 1997; Gupta et al. 1996; Kumar/Divakar 1999). All elasticities between 0 and -1 and the two positive values stem from the study of Kumar/Divakar (1999). Interestingly, the price elasticity estimations for the same brand size are rather different in the studies.

The popular brand size of Jiff 18oz has a price elasticity close to 0.00 (City 1: -0.02, City 2: -0.01) as measured by Kumar/Divakar (1999), whereas the elasticity is around -2.50 (-2.62, -2.53, -2.55 depending on model specification) in the study of Gupta et al. (1996). This example demonstrates once more that it is hard to generalize price elasticities even for a specific brand size. The mean is -1.37 (median -0.83, SD = 1.32).

Figure 4-8: Frequency Distribution of Price Elasticities – Peanut Butter

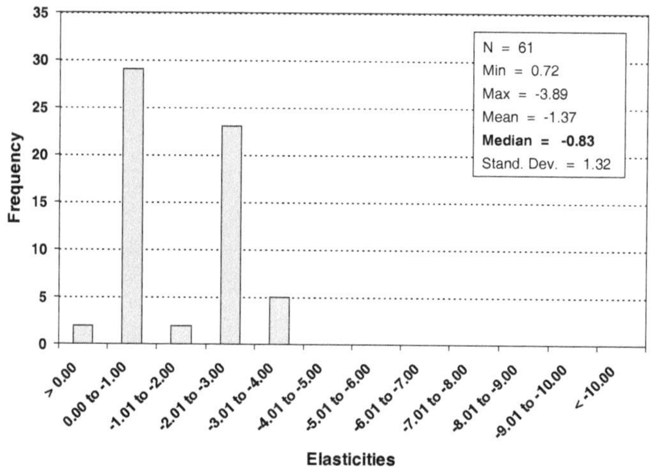

N = 61
Min = 0.72
Max = -3.89
Mean = -1.37
Median = -0.83
Stand. Dev. = 1.32

The distribution pattern for coffee (figure 4-9) is rather broad compared to the other product categories. The 46 price elasticities range from 0.12 to -14.80, the mean is -3.54 (median = -2.70, SD = 2.94). The data stem from four studies (Cooper 1988; Guadagni/Little 1983; Kalyanam 1996; Krishnamurthi/Raj 1991).

High purchase volumes when the product is on sale are typical for this product category and could be a reason for the relatively high price elasticities. Especially, branded products maintain a high shelf price and generate the majority of the sales through price promotions. This leads to high price elasticities for the main brands. The popular brand Chock full o' Nuts has the largest price elasticity (-4.71) in the study of Cooper (1988). This brand is characterized by a high self price and frequent price promotions, 80% of its sales are generated during price promotions (Cooper 1988). The second largest price elasticity in this study (-4.37) is found by the market leader Folgers. The private labels compete primarily on price but with an everyday-low-price strategy, they do not generate enough variation in price to achieve the price elasticities of the branded products (Cooper 1988, p. 714).

However, the study of Krishnamurthi/Raj (1991) finds higher price elasticities for Folgers coffee (loyal customers: -2.70; non-loyal customers: -8.80) than for Chock full o' Nuts coffee (loyal customers: -2.00; non-loyal customers: -6.60), the data also shows higher price elasticities for non-loyal customers than loyal to the brand customers. The remaining two studies of the coffee data set (Guadagni/Little 1983; Kalyanam 1996) do not reveal the brands analyzed, thus a comparison across studies cannot be performed.

Figure 4-9: Frequency Distribution of Price Elasticities – Coffee

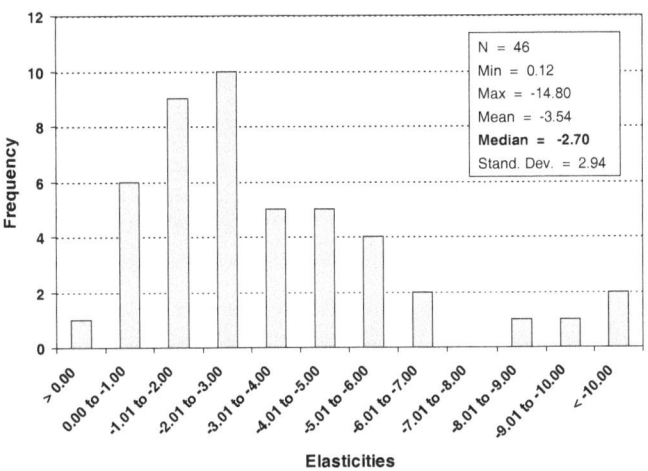

To demonstrate the broad range of research results and differences in price elasticity magnitudes, an additional study on price elasticities for coffee (Bucklin/Srinivasan 1991) is shown in figure 4-10. The data is not included in the overall analysis since the data is not obtained from actual purchases, but self-explicated preference measurement for a 20% price cut in a telephone survey. The frequency distribution and descriptive values illustrate the wide spread of price elasticities within one product category. The mean of the 18 price elasticities is -11.64 (median = -11.15, SD = 5.86). 61.1% of the price elasticities are higher in magnitude than 10.0. The largest price elasticity measured is -26.10, since the respondents were asked about a 20% price cut, the volume would increase by more than 50% (52.2%). This shows a very different picture than the distribution in figure 4-9 and therefore underscores the difficulty to generalize results on price elasticities for a product category.

One reason for lower price elasticities in the panel scanner data could be that consumers' price sensitivities are underestimated without out of stock information. 5%-20% larger price elasticities are measured when out of stock information is accounted for. Out of stock incidences occur frequently when a product is on sale, consumers have to buy another product or not buy. Without the out of stock information, researchers could infer that consumers are not price sensitive, the underestimation of the price elasticities is a result of the absence of out of stock information (Che/Chen/Chen 2012, p. 508).

Figure 4-10: Frequency Distribution of Price Elasticities –
Coffee: Preference Measurement

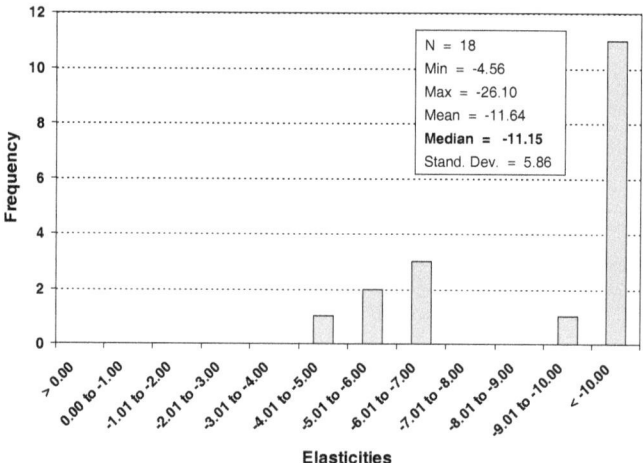

Source: adapted from Bucklin/Srinivasan 1991

The shampoo category (figure 4-11) has a mean price elasticity of -1.21 (median = -1.25, SD = 0.70). The frequency distribution is rather narrow; two thirds of all observations are between -1 and -2. It also might play a role that all observations are derived from one study (Chintagunta 2001). Actually, the study examines only 3 brands but uses 13 different model specifications leading to 39 price elasticities that were all included in the meta-analysis of Bijmolt/Van Heerde/Pieters (2005).

Chapter 4: Magnitude of Price Elasticity

Figure 4-11: Frequency Distribution of Price Elasticities – Shampoo

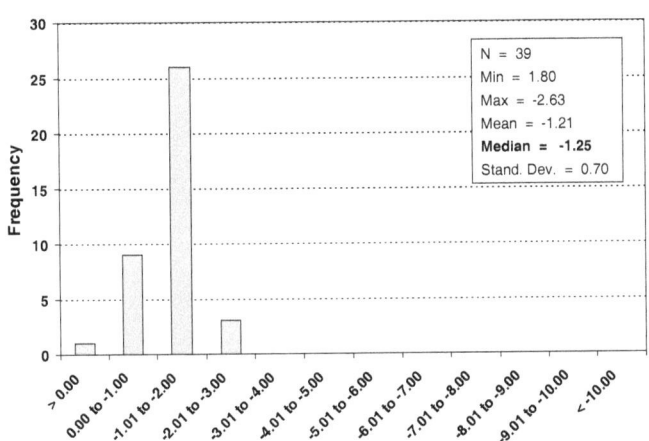

The last two examples are tuna (figure 4-12) and saltine crackers (figure 4-13). Both product categories have a sample size of approximately 30 (33 and 30 respectively) and stem from a similar amount of studies (4 and 3 respectively). The distribution patterns, however, are very different.

The price elasticity distribution is much broader and less strongly peaked in the tuna category. The elasticities range from -1.44 to -8.59 with a mean of -4.00 (median = -3.74, SD = 1.80).

Chapter 4: Magnitude of Price Elasticity

Figure 4-12: Frequency Distribution of Price Elasticities – Tuna

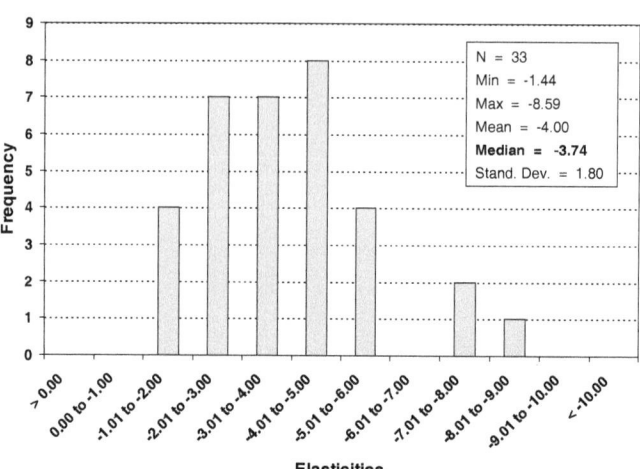

The elasticities for saltine crackers lie between -0.28 and -3.07. The distribution is much narrower and strongly peaked; more than half of the observations (56.7%) lie between -1 and -2. The mean is -1.44 (median = -1.26, SD = 0.70).

Chapter 4: Magnitude of Price Elasticity   65

Figure 4-13: Frequency Distribution of Price Elasticities – Saltine Crackers

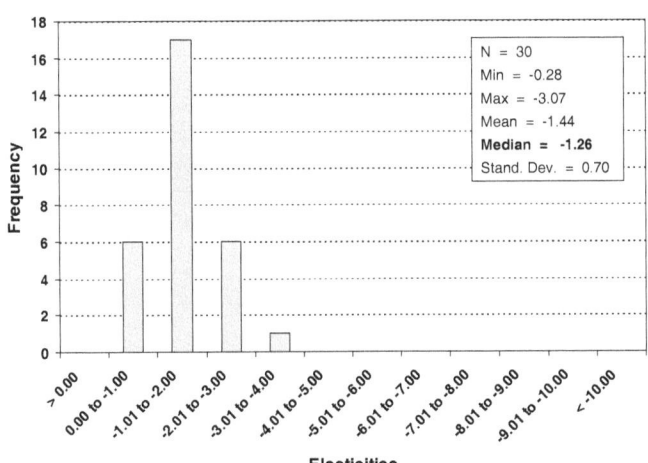

Overall, the examples of the frequency distributions and descriptive values demonstrate the variety of distribution pattern and descriptive values for the various product categories which belong almost exclusively to the fast moving consumer goods market. Furthermore, detailed information on elasticities for specific products and brand sizes illustrate the range in magnitude and that it is not feasible to generalize results.

The overview in table 4-6 compares the descriptive values. First, the overall mean of all price elasticities in the database is listed and then the product categories are listed in order of decreasing magnitude of the mean price elasticity.

Table 4-6: Descriptive Values for Selected Product Categories

Product Category	# of Price Elasticities	Mean	Median	Standard Deviation
All	863	-2.51	-2.21	1.81
Bathroom Tissues	66	-4.09	-4.03	1.60
Tuna	33	-4.00	-3.74	1.80
Coffee	46	-3.54	-2.70	2.94
Margarine	62	-3.26	-3.29	1.67
Ketchup	80	-3.03	-2.85	1.42
Yogurt	64	-2.75	-2.28	2.03
Automotive Hard Parts	66	-2.22	-1.71	1.83
Laundry Detergent	127	-1.93	-1.81	0.78
Saltine Crackers	30	-1.44	-1.26	0.70
Peanut Butter	61	-1.37	-0.83	1.32
Shampoo	39	-1.21	-1.25	0.70

The elasticities are highest for bathroom tissues and tuna (-4.09, -4.00), followed by coffee (-3.54). One reason for the high price elasticities could be the frequent use in promotions. At least for coffee it is proven that high purchase volumes are typical when the product is on sale (Bucklin/Srinivasan 1991; Cooper 1988; Kalyanam 1996). The items are stockable, but that is also the case for the product with the lowest price elasticity – shampoo (-1.21). The median is smaller than the mean for most product categories. Setting the standard deviation in relation to the absolute mean, a normalized measure of the frequency distribution is derived; the resulting coefficients of variation range between 39% for bathroom tissues and 96% for peanut butter.

Looking at price elasticities that indicate a potential to raise prices due to inelastic demand and therefore to increase revenue, there is no potential indicated for laundry detergent, ketchup, bathroom tissues and tuna. A low potential is indicated for yogurt and margarine with 13-15% of the observations lying in the inelastic part of price response curve. A higher potential to increase prices due to a larger percentage of observations being price inelastic can be found for saltine crackers (20%), shampoo (25%) and automotive hard parts (33%). Also peanut butter shows a high potential to increase revenue with a price increase for 50% of the cases; but this situation has to be carefully evaluated since, as mentioned before, the peanut butter Jiff has an elasticity close to 0 in one study (Kumar/Divakar 1999) and in the other study an elasticity of -2.50 (Gupta et al. 1996).

Overall the number of studies and the number of products/brands in relation to the number of models used in one study to determine the price elasticity should not be overlooked in the product categories in order not to jump to conclusions too quickly. One needs to take into account whether the data is obtained from one study with a small number of brands and large number of different models or obtained from a larger number of different studies. The price elasticity estimates can have quite diverse outcomes. It is questionable that there is something like one true price elasticity for one product. Looking at the data it seems that generalizations are difficult to make and the price elasticities have to be evaluated in relation to other price elasticities ideally being calculated with the same estimation method.

4.2 Insights from Consulting Projects

To gain insights from the consulting project data, price elasticities were calculated for each project as described in chapter 3.2.1 and detailed information on the product were recorded. This enabled the author to show a general overview of the price elasticity data (chapter 4.2.1) as well as a more specific analysis by product category (chapter 4.2.2). As the consulting project data provided the opportunity to create a database from a single data source, the analysis of this data is conducted separately from the academic data to value this circumstance.

The same calculation methodology was used for each product and thus one price elasticity estimate per product was obtained. A single price elasticity for each product obtained with the same estimation method provides an enhancement to the comparability of the products and industries.

4.2.1 Overview and Analysis of Price Elasticity Data

In the consulting project data set, the focus is on price elasticity cases for which information on both directions of price changes, price increase and price decrease, could be obtained. Figure 4-14 shows the frequency distribution for the average price elasticities, i.e. the mean of the price elasticity for the price increase and for the price decrease of 10%. The sample consists of 386 price elasticities; the minimum is 0 and the maximum -7.50. The mean is -1.73 (median = -1.29, SD = 1.33). Roughly a third (35.0%) of the elasticities lies between 0 and -1, and another third (33.7%) between -1 and -2. There is no positive price elasticity and the range is from 0 to -7.50. The mean -1.73 is substantially lower than the mean of -2.62 reported by Bijmolt/Van Heerde/Pieters (2005) but close to the mean of -1.76 reported by Tellis (1988).

Figure 4-14: Frequency Distribution of Price Elasticities – Average Price Elasticity of Price Increase of 10% and Price Decrease of 10%

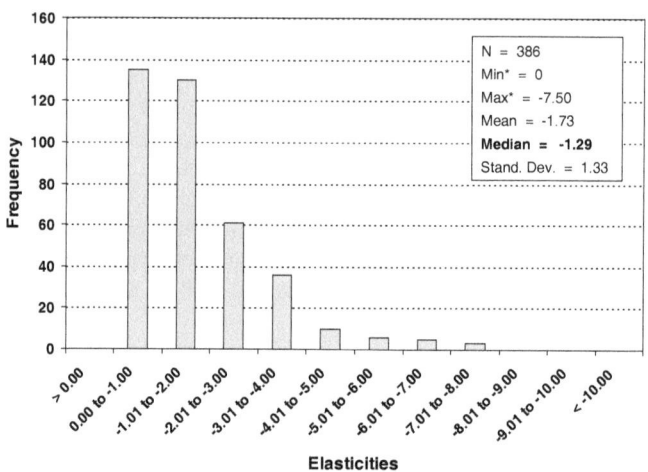

For the price decrease of 10% the price elasticities are more skewed towards 0 (figure 4-15). The mean is -1.62, the median of -1.07 (SD 1.62) demonstrates this skewness as well. 45.9% of all elasticities are between 0 and -1, and 28% between -1 and -2. Two values are positive and the price elasticities range from 2.86 to -11.32.

Chapter 4: Magnitude of Price Elasticity 69

Figure 4-15: Frequency Distribution of Price Elasticities – Price Decrease of 10%

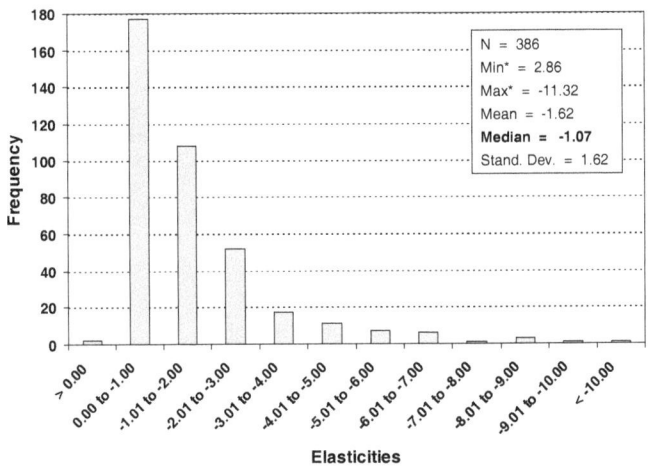

Figure 4-16 shows the price elasticities for a price increase of 10%. The mean is -1.84 (median = -1.50, SD = 1.40). The distribution pattern is similar to the one of the average price elasticity. About a third (32.1%) of the elasticities lies between 0 and -1, and another third (33.7%) between -1 and -2. The price elasticities range from 0.27 to -8.09.

Figure 4-16: Frequency Distribution of Price Elasticities – Price Increase of 10%

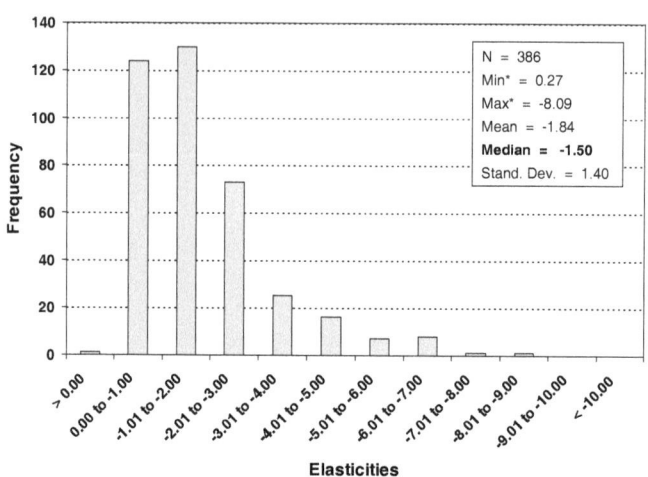

There is a significant difference between the means of -1.62 for the price decrease and -1.84 for the price increase. The mean price elasticity for the price decrease is significantly lower than the price elasticity for the price increase (p < 0.01). The weaker reaction to price decreases is also indicated by the lower median of -1.07 vs. -1.50 for the increase. The asymmetry in price elasticity is supported by (Bidwell/Wang/Zona 1995; Hanssens/Parsons/Schultz 2001, p. 334; Monroe 2003, p. 149; Putler 1992, p. 304). Classical microeconomic theory does not imply asymmetric responses to price changes but predicts that consumers react to a small price increase in much the same way as they do to a small price decrease. However, consumers tend to react more quickly and strongly when prices are raised than when prices are reduced (Bidwell/Wang/Zona 1995).

This is consistent with behavioral research concerning how consumers perceive gains and losses. Prospect theory of asymmetric gains and losses says that consumers tend to experience losses much more intensely than they value gains (Kahneman/Tversky 1979; Thaler 1985). Paying a price can be interpreted as a loss since it reduces the consumer's available budget. Given a reference price, a price increase can be interpreted as a loss and a price reduction as a gain for the consumer. In the research at hand the reference price is the price anchor (the current price of the product) used to evaluate price changes of 10%. Therefore, according to prospect theory a price increase of the same amount is valued more negatively than a price reduction of the same amount is valued positively. This leads to higher price elasticities for price

Chapter 4: Magnitude of Price Elasticity

increases than price reductions. The phenomenon that losses loom larger than gains is known as loss aversion.

However, not all research is in line with this reasoning. Pauwels/Srinivasan/Franses (2007) find increased price elasticities for gains (price decreases) and lower price elasticities for losses (price increases). They explain it with consumers waiting for deals (Mela et al. 1997) and then reacting strongly. The research at hand, however, does not analyze deals but long-term price changes. Novemsky/Kahneman (2005) and Ariely/Huber/Wertenbroch (2005) define limits to loss aversion and suggest moderators such as emotional attachment on loss aversion. In addition, Chen et al. (2012, p. 64) state that a price reduction might be perceived as a reduction in loss rather than a gain, which could be another influencing factor on the differing research results.

For all three price elasticity calculation methods (average, price decrease, price increase) roughly 85% of all observations lie between 0 and -3 (84.5%, 87.3%, 84.7%). About a third of all price elasticities lie in the inelastic area between 0 and -1, which indicates a potential to raise prices in order to optimize revenue and profits.

As mentioned before, only the price elasticities for which both directions of price changes could be measured in the consulting project are the basis of the analysis and comprise 386 price elasticity cases. To complement the data, the frequency distributions for all available cases are shown in figure 4-17 for a price decrease and figure 4-18 for a price increase.

For a price decrease, there are 415 price elasticity cases ranging from 2.86 to -11.32. The mean is -1.59 and the median is -1.07 (SD 1.58).

***Figure 4-17: Frequency Distribution of Price Elasticities –
Price Decrease of 10 % (all Data)***

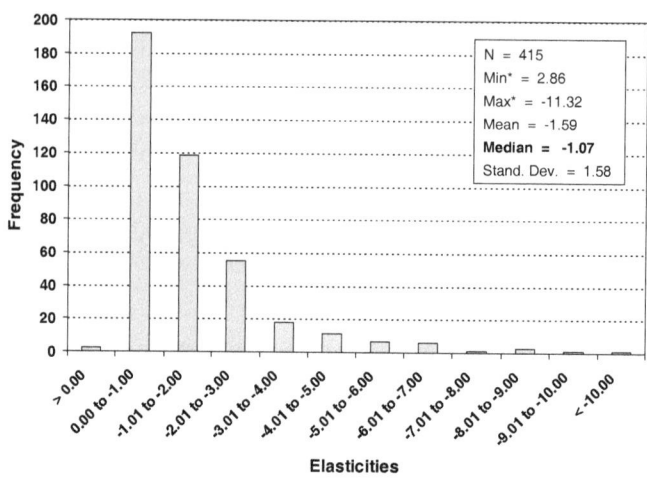

For a price increase (figure 4-18), there are 409 price elasticity cases ranging from 0.27 to -9.12. The mean is -1.87 and the median is -1.55 (SD 1.43).

Chapter 4: Magnitude of Price Elasticity 73

*Figure 4-18: Frequency Distribution of Price Elasticities –
Price Increase of 10% (all Data)*

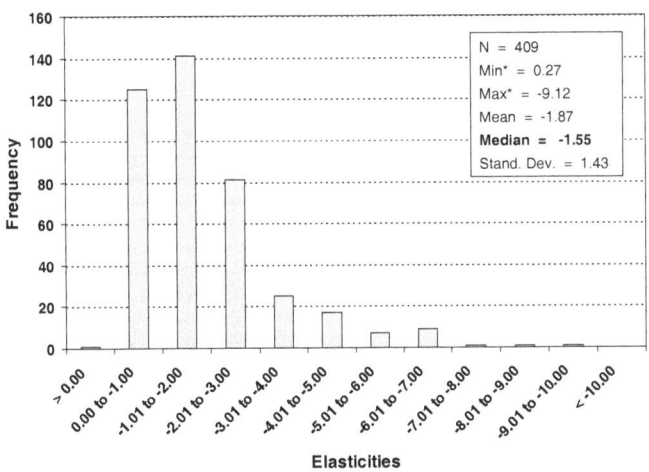

Also comparing the means and medians for a price decrease and a price increase, there are stronger reactions for the price increase, which can be illustrated comparing the median of -1.55 for the increase and -1.07 for the decrease. The median is considered to be a more meaningful indicator than the mean since it is less influenced by outliers (Simon/Fassnacht 2009, p. 104). However, comparing the magnitude of the price elasticities for a price increase and a price decrease, it is recommended to assess the reactions for the same product cases to ensure comparability. The data set contains 386 price elasticity cases that have both data for a price increase and a price decrease. Thus the strength of the price elasticity being influenced by the fact that the product and industry compositions differ in the data set is not an issue. Therefore, the comparison of the means for this data set is more reliable. As mentioned earlier, there is a significant difference ($p < 0.01$) in the means of -1.62 for the price decrease and -1.84 for the price increase. The weaker reaction to a price decrease is also indicated by the lower median of -1.07 vs. -1.50 for the price increase.

Table 4-7 shows an overview of the products representing the price elasticity database. It consists of a vast variety of products and covers a broad range of industries. The data includes fast moving consumer goods which is, as shown in chapter 4.1.1, the main basis of previous analyses in academic research. However, it also includes consumer durables such as tableware and glassware, services like mail and parcel services as well as medical products such as medication and medical devices. Selected examples from the business to business industry include electrodes, construction cranes, films for printing as well as rotorspinning and die casting machines.

Table 4-7: Price Elasticities by Product Category

Product Category	Total Number of PE	Product Category	Total Number of PE
food & beverages	58	copper crim fitting	6
automobile	41	technology	5
automotive tires	26	tunnel toll	5
parcel services	25	cranes and crane parts	4
self-adhesive products	19	insurance	4
tableware	18	bed mattress	3
home food packaging products	17	label printing press	3
cosmetics	15	mail services	3
medication	15	telecommunication packages	3
groceries	14	airway cargo	2
medical device	12	dermal filler	2
respirators	12	newspaper subscription	2
electrodes	11	rotorspinning	2
financial services	10	service contracts for elevators	2
glassware	10	welding equipment	2
telecommunication package	9	die casting machine	1
film for printing/graphics	8	hammer drill	1
newspaper advertisement	8	ski rental	1
automotive spare parts	6	train transportation	1
		Total = 386	

4.2.2 Analysis of Selected Product Categories

In this chapter, price elasticity distributions for selected product categories are presented. The product categories and industries are analyzed in order of decreasing sample size. The automotive category has the largest sample size with 73 cases; it contains data on automobiles, automotive tires and automotive spare parts. The frequency distribution and descriptive values are presented in figure 4-19. The mean price elasticity is -2.02 (median = -2.00, SD = 1.16). 72.6% of all price elasticities lie between -1 and -3.

Chapter 4: Magnitude of Price Elasticity 75

Figure 4-19: Frequency Distribution of Price Elasticities – Automotive

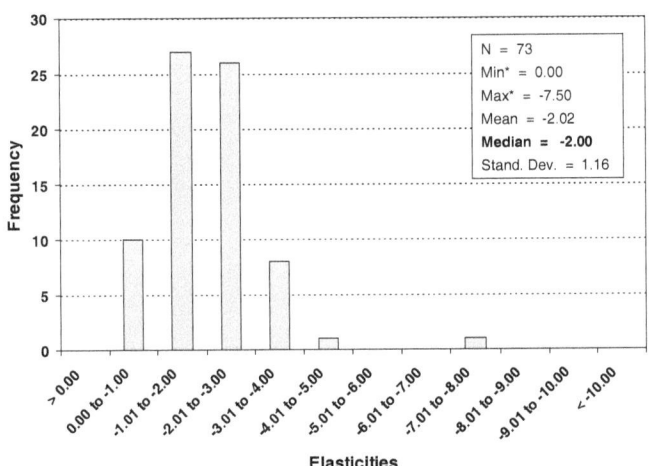

Looking only at automobiles within the automotive industry, the descriptive values do not change very much. The mean of -1.96 for automobiles only is rather close to the mean of -2.02 for the whole automotive industry (other descriptive values for automotive only: median = -1.80, SD = 1.42, n = 41).

The next category is the backbone of previous academic research results – fast moving consumer goods. Comparing the product range included in this category with what was included in the academic data set (cf. chapter 4.1.1, table 4-1), shows that the database of this research expands the previously studied product range. Besides food (frozen pizza, eggs, desserts, chocolate bars, etc.) and beverages (alcoholic and soft-drinks), the data includes cosmetics (facial cream, mascara, nail polish, etc.), home food packaging products (freezer bags, aluminum foil, etc), feminine hygiene products, dog food and more. Figure 4-20 shows the frequency distribution and the descriptive values for the 60 price elasticities in this category. The mean is -1.14 (median = -0.85, SD = 1.04) and therefore substantially lower than the mean (-2.51) obtained from the adjusted database of Bijmolt/Van Heerde/Pieters (2005, cf. figure 4-1). 60.0% of all cases lie between 0 and -1. This indicates that there is substantial potential to raise prices in this product category.

Figure 4-20: Frequency Distribution of Price Elasticities – Fast Moving Consumer Goods

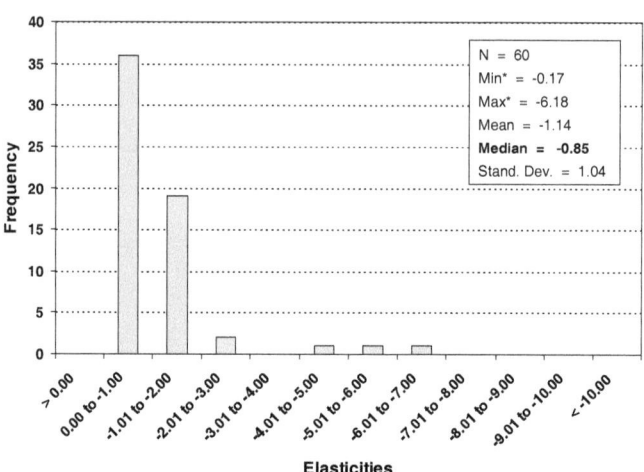

The next product category – industrial goods – represents business to business transactions. The products include, for example, a label printing press, cranes and crane parts, copper crim fittings and self-adhesive products. The frequency distribution of the 56 price elasticities is displayed in figure 4-21. The mean is -1.54 (median = -1.29, SD = 1.03). Slightly more than a third (35.7%) of the elasticities lie between 0 and -1, and the same amount lies between -1 and -2.

Chapter 4: Magnitude of Price Elasticity

Figure 4-21: Frequency Distribution of Price Elasticities – Industrial Goods

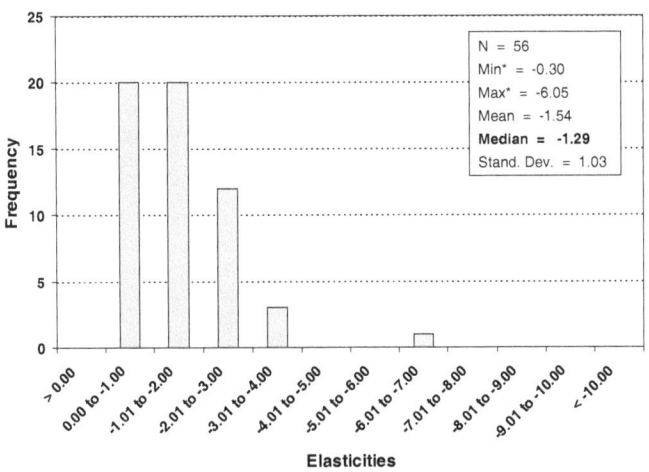

The next category represents the away from home food market (n = 44), especially snacks and fast food while travelling. This market is represented by stores and fast food restaurants at train stations, airports and gas and rest stations on freeways. The frequency distribution and the descriptive values are shown in figure 4-22. The mean of -3.41 is higher than in the previous categories. The median is -3.26 (SD = 1.74). The peak is between -3 and -4 (29.5% of cases), the standard deviation indicates a broader distribution compared to the other product categories.

Figure 4-22: Frequency Distribution of Price Elasticities – Away from Home Food

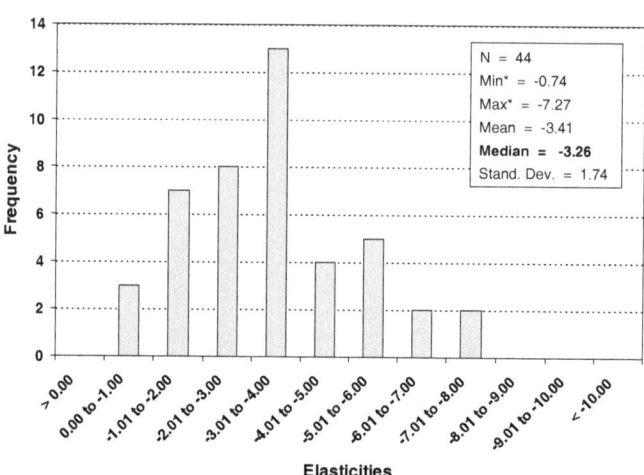

This product category is characterized by high price elasticities. Numerous products are overpriced in this sample, this means the prices are above the profit-optimal price level. For many products strong price reactions exist in one direction of the price change, either for the price decrease or increase depending if the price threshold is crossed increasing the price or lowering the price. To give an example, one product has a price elasticity of -8.09 when the threshold of 5 German Marks, which used to be the largest coin, is crossed increasing the price; the elasticity is no more than -3.31 for the price reduction. In some cases, the product price is still above the optimal price level when the price is decreased by 10% but there is a strong change in buying volume. For current prices that are close to the profit-optimal price, price changes do not cause such strong price reactions. For example, a bread roll with meatloaf is priced at 13.00 DM, a 10% price decrease leads to a price of 11.70 DM, which is still above the profit-optimal price that is less than 10.00 DM and this price change leads to a price elasticity of -11.32. For the price increase the price elasticity is only -3.03 since it is already substantially above the optimum, so a further increase does not change the already low demand that much. The average price elasticity is -7.17 in this case. The high price elasticities are also in line with a survey that revealed that 76% of consumers believe that prices at rest stops in Germany are not appropriate, especially the prices for meals are considered to be far too high (N.N. 2010, N.N. 2011a, N.N. 2011b).

Chapter 4: Magnitude of Price Elasticity 79

The next category consists of logistics (n = 36), for example postal and cargo services. The frequency distribution is displayed in figure 4-23. Again this distribution is not bell curved but strongly peaked between 0 and -1 with 38.9% of all observations in the inelastic range. The mean is -1.62 (median = -1.41, SD = 1.26).

Figure 4-23: Frequency Distribution of Price Elasticities – Logistics

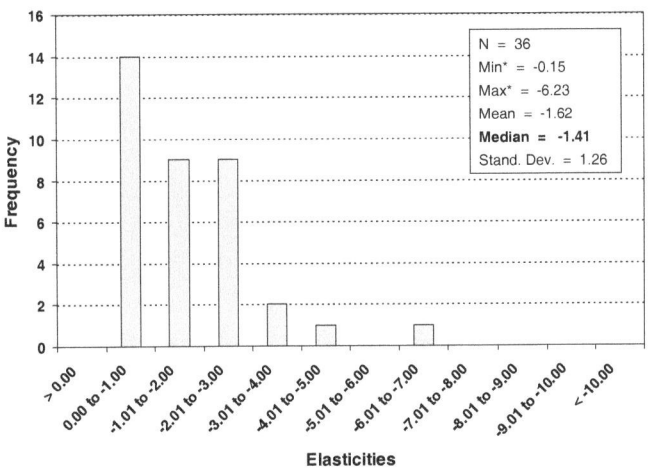

As mentioned, there are no consumer durables in the meta-analysis of Bijmolt/Van Heerde/Pieters (2005) looking at the data for the last 25 years. In contrast to this, the database of the research project at hand includes 32 cases of consumer durables, comprising bed mattresses, tableware, glassware and kitchen electronics. The frequency distribution and the descriptive values of this category are displayed in figure 4-24. The price elasticities range from -0.72 to -4.21. The mean is -1.40 (median = -1.12, SD = 0.86). More than half of all cases (53.1%) lie between -1 and -2. There are no elasticities in the range of -2 and -3. The highest price elasticity of -4.21 stems from a price increase above an important price threshold of 3,000 USD, thus led to a price elasticity of -6.62. For the price reduction of 10%, however, no major price threshold was crossed and led to a price elasticity of -1.80. The average price elasticity is therefore -4.21. This is similar for the other outliers in this elasticity distribution. The example illustrates that price elasticity calculations should always be evaluated in the context of the market and pricing environment.

Chapter 4: Magnitude of Price Elasticity

Figure 4-24: Frequency Distribution of Price Elasticities – Consumer Durables

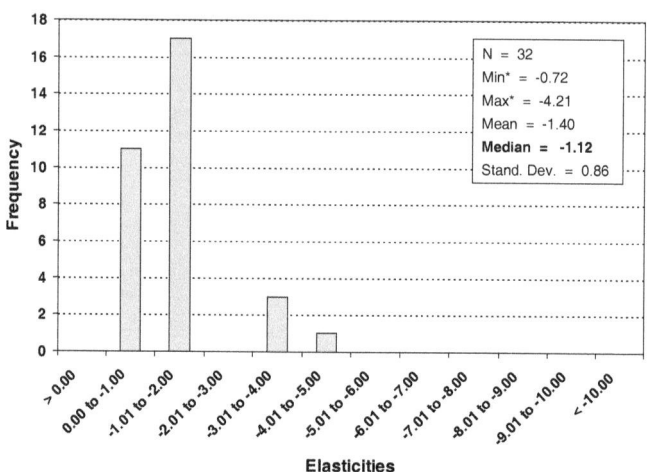

The frequency distribution and descriptive values for pharmaceuticals and medical technology are displayed in figure 4-25. The 32 products cover a wide range of treatments. For example, the data includes medications for wide-spread diseases like hypertension, dyslipidemia, pulmonary disease and gastro reflux disease. However, also data on dermal fillers for cosmetic treatments and plastic surgery, cancer medication as well as medical devices for gynecological surgeries are included. The price elasticity range is relatively narrow from -0.82 to -3.19. Most values (93.1%) are between 0 and -2. There are only two values larger in magnitude than -2. These relatively high price elasticities of -3.19 and -3.06 are derived from a third generation product in a market with available treatment options that are both good and affordable. The disease is not life threatening and a very good treatment option is already available as a generic medication. The manufacturer released a slightly improved second-generation product as a follow-up to the original medication and now was testing a third generation product with a slightly different molecule. Physicians did not see much value over the existing products. Therefore, their prescribing behavior exhibited a high price elasticity.

The mean of the price elasticities for pharmaceuticals and medical technology is -1.33 (median = -1.09, SD = 0.58), which is close to Tellis' (1988, p. 334) findings of a mean = -1.12 (n = 52). Bijmolt/Van Heerde/Pieters (2005) analysis does not include any pharmaceuticals for the last 25 years. A more recent study finds price elasticities

Chapter 4: Magnitude of Price Elasticity   81

in the range of -2.37 to -3.06 for antihistamines (Narayanan/Desiraju/Chintagunta 2004, p. 98).

Figure 4-25: **Frequency Distribution of Price Elasticities –
Pharmaceuticals and Medical Technology**

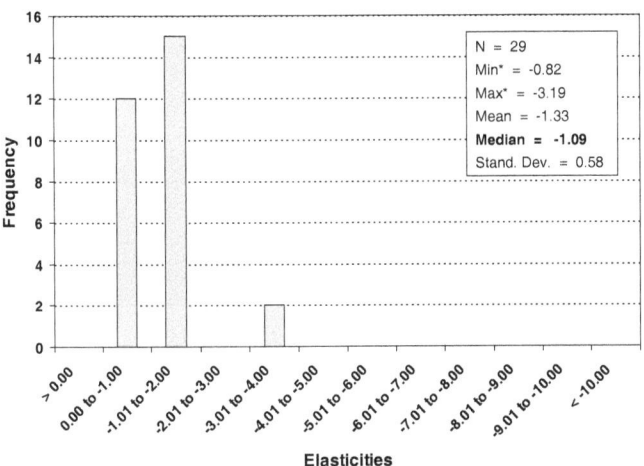

Table 4-8 shows an overview of the descriptive values for the selected product categories in order of decreasing magnitude of price elasticity. The overall mean is -1.73 and the median of -1.29 is lower than the mean (SD = 1.33). The median is lower than the mean in all categories.

The highest mean by far of -3.41 can be found in the away from home food market. The lowest mean is to be found in the fast moving consumer goods category. The coefficient of variation (SD/mean) is highest for the fast moving consumer goods (91%), which in turn shows the lowest price elasticity (-1.14). It is unusual that permanent price changes are analyzed in the fast moving consumer market. The higher price elasticities observed in the academic data (cf. chapter 4.1.1 and 4.1.2) might be influenced by the fact that these result primarily from temporary price changes such as price promotions and consumers adjust their demand more strongly when price promotions are analyzed. The rather low price elasticities for pharmaceuticals and medical technology are in line with previous research (Tellis 1988). The standard deviation in relation to the mean shows the lowest variation in this category (44%). In the other product categories, the mean ranges between -1.40 and -2.02.

Table 4-8: Descriptive Values for Selected Product Categories

Product Category	# of Price Elasticities	Mean	Median	Standard Deviation
All	386	-1.73	-1.29	1.33
Away from Home Food	44	-3.41	-3.26	1.74
Automotive	73	-2.02	-2.00	1.16
Logistics	36	-1.62	-1.41	1.26
Industrial Goods	56	-1.54	-1.29	1.03
Consumer Durables	32	-1.40	-1.12	0.86
Pharma and Medtech	29	-1.33	-1.09	0.58
FMCG	60	-1.14	-0.85	1.04

4.3 Summary

In this chapter, the magnitude of price elasticity was analyzed. First, the previously created academic data set was examined and second, the consulting project data was examined. In table 4-9, the overall mean and other descriptive values of these data sets are compared with the results of the previous meta-analyses of Bijmolt/Van Heerde/Pieters (2005) and Tellis (1988).

Chapter 4: Magnitude of Price Elasticity

Table 4-9: Comparison of Descriptive Values for Selected Data Sources

Data Source	# of Price Elasticities	Mean	Median	Standard Deviation
Academic Data Set	863	-2.51	-2.21	1.81
Consulting Project Data	386	-1.73	-1.29	1.33
Bijmolt/Van Heerde/ Pieters (2005)	1851	-2.62	-2.22	2.21
Tellis (1988)	367	-1.76	not reported	1.74

The mean price elasticity of the academic data set is -2.51 with a median of -2.21 and a standard deviation of 1.81. This is very much in line with Bijmolt/Van Heerde/Pieters' (2005) mean of -2.62 (median = -2.22, SD = 2.21). The consulting project data mean is at -1.73, substantially lower (median = -1.29, SD = 1.33) but close to Tellis' (1988) mean of -1.76 (SD = 1.74). Tellis' (1988) database contains more diverse data with a larger percentage of durables and pharmaceuticals than Bijmolt/Van Heerde/Pieters (2005) and the academic data set that contains no durables and pharmaceuticals for the last 25 years of research. This could also contribute to the fact that the descriptive values of the consulting project data are closer to Tellis' (1988) values.

The consulting project data provided the opportunity to compare price elasticities for a price increase with price elasticities for a price decrease (table 4-10). The mean of the price decrease (-1.62) is significantly lower (p < 0.01) than that of the price increase (-1.84). The median illustrates the stronger reaction to price increases even better (-1.50 vs. -1.07) since it is less influenced by outliers.

Table 4-10: Comparison of Descriptive Values for a Price Decrease vs. a Price Increase

Data Source: Consulting Project Data	# of Price Elasticities	Mean	Median	Standard Deviation
Price Decrease	386	-1.62	-1.07	1.62
Price Increase	386	-1.84	-1.50	1.40

To expand the knowledge on price elasticities, the academic data set and the consulting project data were analyzed by product category. The actual composition of the products included in the meta-analysis of Bijmolt/Van Heerde/Pieters (2005) was not provided. Their products were categorized as groceries with high and low stockpiling ability and durables. In order to find out more about the products included in the data set, a review of each price elasticity study, contained in the newly created academic data set, was conducted. An overview of the product categories revealed that a third of the academic data consists of price elasticities for laundry detergent, ketchup and bathroom tissues, adding yogurt, margarine, peanut butter and automotive hard parts (automotive hard parts were not included in the analysis of Bijmolt/Van Heerde/Pieters 2005 since the study of Mantrala et al. was published in 2006) already covers more than 60% of all data (cf. table 4-1). This illustrates to what extent current research results are derived from a rather narrow group of product categories.

The data on price elasticities was expanded by adding more diverse data derived from consulting projects (cf. table 4-7). The current knowledge on price elasticities is enhanced by providing frequency distributions and descriptive analyses for all product categories with a minimum sample size of 30 cases for the academic data and 29 cases for the consulting project data (cf. chapter 4.1.2 and chapter 4.2.2). The descriptive values are compared in a tabular overview at the end of each chapter (cf. table 4-6 and table 4-8).

The price elasticities have rather diverse distribution patterns. Some are more strongly peaked, others more equally distributed or bell-curved. Overall, all these figures are explicitly shown to illustrate the price elasticity range and the various distribution patterns visually. Looking at these figures, the wide variety of distribution patterns and differences between product categories become evident. When there is an opportunity to go deeper into the data analysis, brands and brand sizes are also compared across studies. There is a broad range of price elasticities within product categories, within brands and even within brand sizes, which makes generalizations on price elasticities difficult. Examples that illustrate this can be found in the deeper analyses of the ketchup category (table 4-3) and the yogurt category (table 4-4) that for the same brand size substantially different price elasticities are measured. The price elasticities measured for Heinz ketchup 28 oz varies widely across studies, with the highest magnitude being 5 times larger than the smallest price elasticity measured. For Dannon yogurt the 29 price elasticities found range from -0.82 to -10.50, which also illustrate that no generalization concerning the price elasticity is feasible. The author also explains in various examples, why some price elasticities are substantially higher than others within the sample, e.g. when price thresholds are crossed or the price is too far away from the optimal price point (e.g. in the away from home food market). All of this illustrates the broad variety in the magnitude of price elasticity for various product categories.

5 Determinants of Price Elasticity

In a special issue of Marketing Science on empirical generalizations in marketing, Bass/Wind (1995, p. 1) indicate that "science is a process in which data and theory interact leading to generalized explanations of disparate types of phenomena". They also note that in the last decades few generalized explanations have been produced. One reason for the lack of progress is that there has not been enough empirical research prior to theory development and too much research which is restricted to theory prior to empirical research (Ehrenberg 1995). A common method for reporting research is to formulate hypotheses on the basis of known theory or prior research – theoretical generalization (Blair/Zinkhan 2006, p. 5 f.; Ehrenberg 1995). In understanding the determinants of price elasticities there is little established theory to draw on and previous research has often produced conflicting results (Danaher/Brodie 2000, p. 919; Bijmolt/van Heerde/Pieters 2005; Tellis 1988; cf. also chapter 2.2.2).

The requirement of managerial relevance in generalization is also debated in research discussions (Bass/Wind 1995, p. 2). Generalization is also created through replication. If findings are considered to be central, other researchers will elaborate on them. In the process, if the finding is a coincidence of the sample or else not robust, it will become apparent (Blair/Zinkhan 2006, p. 6). For these reasons, a more explorative empirical approach seems appropriate. Bijmolt/van Heerde/Pieters (2005, p. 154) ask specifically for research that extends the scope and knowledge of price elasticities beyond the range of the current meta-analytic design. Also Blair and Zinkhan (2006) stress the importance to be aware of the cumulative research record as well as to build and favor research that adds higher value through diversity. In addition, they also state concerns regarding the extent to which data used in a research project reflect a broader population (Blair/Zinkhan 2006, p. 4). To extend the range beyond previous meta-analytic designs, a major focus in this research project is on incorporating diverse variables to explore additional determinants of price elasticities. The extension of knowledge about determinants of price elasticities is also addressed by using new data sources, a broader product range, consistent calculation methodology, long-term price changes and regular prices.

As it became evident in chapter 3.1.2, the majority of research articles concentrate on aspects of the research methodology, if they analyze determinants of price elasticity at all. This determinant group is sufficiently covered in the meta-analysis of Bijmolt/Van Heerde/Pieters (2005; cf. table 2-1 in chapter 2.2.2) and further analysis of methodological determinants will provide limited insights. In addition, there is a limited ability to derive insights for managerial implications from these determinants.

The second determinant group used by Bijmolt/Van Heerde/Pieters (2005) is market characteristics. Table 2-2 (chapter 2.2.2) provides an overview of previously analyzed variables and shows areas in which to deepen the analysis. For example the brand ownership was not analyzed by Tellis (1988) and had no significant results in the meta-analysis of Bijmolt/Van Heerde/Pieters (2005). The analysis of product categories was rather limited and other variables like the product life cycle showed conflicting results.

To determine the spectrum of determinants and broaden the existing knowledge about determinants of price elasticities in various environments, the following approach was chosen. First, the academic literature, based on the data set created, was examined. The author explored which determinants were actually examined in the original research and which additional insights could be derived. In the next step, based on the information gathered and previous research results, interviews with experts in the pricing field were conducted to test the influence of the determinants and to investigate additional determinants using open questions. In the subsequent steps the consulting project data was coded and analyzed.

5.1 Insights from Academic Publications

In order to gain insights on the determinants of price elasticity, the 46 academic publications that had been identified as containing price elasticity information according to the selection criteria (cf. chapter 3.1.1 and table 3-1) were analyzed. An explorative article by article analysis was conducted. Hereby, the author did not only examine what previous research focused on but what additional insights could be gained from the data set. During this process, each academic article was looked at to check if determinants were explicitly studied and if yes, which ones were studied. In addition, the data provided in the studies was examined to find additional indicators of other insights and implications to be drawn. Available data on price elasticities was recorded along with information on corresponding market shares, price levels, brands, etc.

Having gathered all information on potentially influencing factors of price elasticity, the information was evaluated and it became apparent that several factors could be used to assess their impact on the magnitude of price elasticity. The information was grouped into the following categories: market share (chapter 5.1.1), competition (chapter 5.1.2), premium positioning and quality (chapter 5.1.3), brand ownership (chapter 5.1.4), direction of price change (chapter 5.1.5) and customer characteristics (chapter 5.1.6). These categories were formed after all studies had been analyzed regarding any information that could lead to insights on influencing factors on the price elasticities. Figure 5-1 shows these categories of determinants as identified in the academic publications.

Chapter 5: Determinants of Price Elasticity

Figure 5-1: Categories of Determinants Identified in Academic Publications

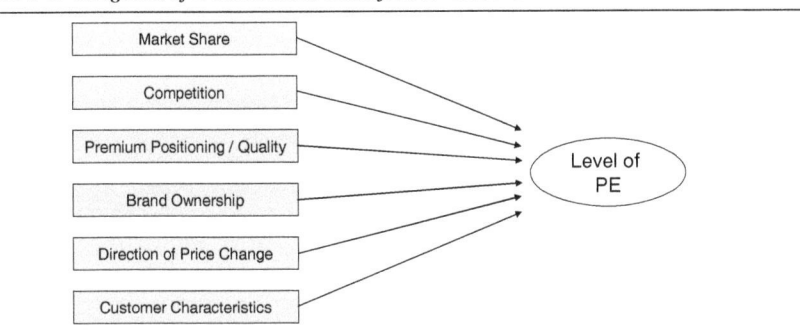

In some instances, the data is complemented with additional information derived from further studies originally not included in the data set, for example because they use a data source other than real purchase data or averages across products and would therefore bias the data analysis on the magnitude of price elasticities.

5.1.1 Market Share

The market share is the first category presented. Due to the mathematical calculation of the price elasticity as the percentage change in quantity in relation to the percentage change in price, it can be assumed that smaller share brands have higher price elasticities than brands with larger market shares.

Out of the 46 publications in the academic data set, nine are suitable to obtain insights on the influence of market share on price elasticities. The relationship between market share and price elasticity was not analyzed in the original studies but for the research project at hand these analyses were performed and correlations were calculated with the data of the studies listed in table 5-1. Three additional studies (Ailawadi/Lehmann/Neslin 2001; Bucklin/Srinivasan 1991; Jedidi/Mela/Gupta 1999) are included since they enable insights into the relationship between market share and price elasticities. They were not included in the analysis of the magnitude of price elasticity as they, for example, use averages across products or are derived from telephone surveys rather than real purchase data. Once again, the fact that the impact on the price elasticity is on the absolute magnitude is stressed. Thus a plus sign in the table indicates an increase in the absolute magnitude of price elasticity. In other words, a plus sign indicates that the higher the market share, the higher the price elasticity. A minus sign indicates a negative impact of the market share on the price elasticity. The higher the market share, the lower the price elasticity.

Looking at table 5-1, the overall picture supports the assumption that large share brands tend to have smaller price elasticities than small share brands.

Table 5-1: Determinant – Market Share

Study	Product Category	Key Results	Relationship with PE
Ailawadi/Lehmann/ Neslin 2001	packaged consumer good	PE large and mid-sized brands < PE small brands	-
Allenby/Rossi 1991	margarine	PE large share brands < PE small share brands	-
Bucklin/Srinivasan 1991	coffee	PE large share brands < PE small share brands	-
Chintagunta/Honore 1996	saltine crackers	PE large share brands < PE small share brands	-
Chintagunta/Jain/ Vilcassim 1991	saltine crackers	PE large share brands < PE small share brands	-
Cooper 1988	coffee	PE large share brands > PE small share brands	+
Guadagni/Little 1983	coffee	PE large share brands < PE small share brands	-
Jedidi/Mela/Gupta 1999	unrevealed consumer packed good	PE large share brands < PE small share brands	-
Kadiyali/Chintagunta/ Vilcassim 2000	orange juice/ tuna	Correlations too low	o
Kamakura/Russell 1989	unrevealed food item	PE large share brands < PE small share brands	-
Mulhern/Williams/ Leone 1998	liquor	Mixed results for different liquor categories	+/-/o

- : negative relationship with the magnitude of PE (correlations range from 0.58 to 1.00)
+: positive relationship with the magnitude of PE (correlations range from -0.63 to -0.76)
o: no or low observable relationship with the magnitude of PE (correlations range from -0.19 to 0.10)

Cooper (1988) is the only study exclusively showing a positive relationship of brand size and price elasticity. The third largest brand (out of 12) has the highest price elasticity (-4.71). This brand is characterized by a high shelf price and frequent price promotions, 80% of its sales are during promotions. The market leader (28.5% market share), whose price is medium high, has the second highest price elasticity (-4.37). The correlation between brand size and price elasticity is -0.76 for the whole data set.

Mixed results are also found for liquor. Some liquor categories feature very prominent brands with high market shares (e.g., Bacardi Rum and Smirnoff Vodka). These brands draw sales away from the other brands when promoted, and for that reason higher market share brands have higher price elasticities (Mulhern/Williams/Leone 1998, p. 441). However, in other liquor categories like Scotch the relationship of market share and price elasticity is reversed.

Chapter 5: Determinants of Price Elasticity   89

Significant differences in price elasticity can be found for the three largest brands in the tuna category. The correlation of market share and price elasticity is -0.19 for tuna and 0.10 for orange juice, indicating no substantial influence on market share (Kadiyali/Chintagunta/Vilcassim 2000).

Chintagunta/Honore (1996) show that smaller share brands have higher price elasticities than larger share brands. The correlations for four brands of saltine crackers range from 0.77 for the probit model to 0.90 for the logit and also the mixture of logit models. The market leader Nabisco has the lowest price elasticity.

Additional insights can be gained from two further publications not part of the original data set due to the aggregation level and data source. The article of Ailawadi/Lehmann/Neslin (2001) explicitly mentions the relationship between brand size and price elasticity. However, the data is not included in the academic data set since averages across categories are used. In two studies more than 100 brands across 24 product categories in the packaged consumer goods are examined. In both studies the price elasticities of small brands are larger than the price elasticities of mid-sized and large brands.

Bucklin/Srinivasan (1991) analyze price elasticities of coffee derived from a self-explicated preference structure model via telephone interviews. The data is displayed in chapter 4.1.2, figure 4-10. Price elasticities for large share brands are smaller than for small share brands. The large price elasticities for the small share brands are attributed to the small base sales volumes for these brands (Bucklin/Srinivasan 1991, p. 65 f.). The correlation of market share and price elasticity is 0.67.

5.1.2 Competition

Another aspect considered in the academic data set is the influence of competition. Table 5-2 shows the studies with various aspects of competition and their influence on the magnitude of price elasticity. The various variables are taken from the original studies; competition is defined and measured in different ways. These variables are comprised and presented under the category competition without making changes to the original variables.

Two studies (Hoch et al. 1995; Montgomery 1997) examine the same four competitive variables as well as consumer variables which will be discussed in chapter 5.1.6. Competitive characteristics from two sources, supermarkets and warehouse/superstore operations are considered. The distance to these stores and the sales volume relative to these competitors (as a proxy for store size) are used as determinants. Two of the four competitive determinants have congruent results in both studies. First, the longer the distance to the nearest warehouse, the lower the price elasticity. Second, the higher the relative sales compared to the warehouse, the higher the price elasticity. For the other competitive variables there are only weak and mixed effects across the various product categories. Overall, "the results suggest that the characteristics of the competitive environment are not all that important as determinants of store price sensitivity" (Hoch et al. 1995, p. 28). Montgomery (1997) comes to a similar conclusion. Competitive characteristics add little explanatory power to the magnitude of price elasticity (Montgomery 1997, p. 321; Hoch et al. 1995, p. 24).

Table 5-2: Determinant – Competition

Study	Product Category	Key Results	Impact on Elasticity
Hoch et al. 1995	packaged consumer goods	▪ avg. distance to the nearest 5 supermarkets: weak and mixed effects on PE	+/-
		▪ distance to nearest warehouse operator: reduces PE	-
		▪ sales volume of store relative to supermarket competitor: weak and mixed effects	+/-
		▪ sales volume of store relative to warehouse competitor: increases PE	+
Montgomery 1997	orange juice	▪ avg. distance to the nearest 5 supermarkets: increases PE, weak effect	+
		▪ distance to nearest warehouse operator: reduces PE, weak effect	-
		▪ sales volume of store relative to supermarket competitor: increases PE, weak effect	+
		▪ sales volume of store relative to warehouse competitor: increases PE, weak effect	+
Mantrala et al. 2006	automotive hard parts	▪ areas with high concentration of own stores: reduces PE	-
Van Heerde/Mela/ Manchanda 2004	pizza	▪ introduction of new, innovative brand increases PE of existing brands	+

- : negative impact on magnitude of PE
+ : positive impact on magnitude of PE

An automotive part retailer that caters to customers who repair and maintain their cars themselves represents a specialty market. In such a market, a high concentration of own stores within a given area reduces the price elasticity (Mantrala et al. 2006). The retailer faces less competition from other retailers due to the concentration of own stores with a similar product pricing. Therefore, the consumers have no real alternatives when prices are increased (Mantrala et al. 2006, p. 599).

The competition within a market is also influenced by the product range available and the market entrance of new products. Van Heerde/Mehla/Manchanda (2004) show that the introduction of a new innovative brand in the pizza market increases the price elasticities of the existing brands. Innovation decreases the brand differentiation for the existing brands and thus the price elasticity increases. Only brands at the extreme ends, high quality, high-priced and low-quality, low priced brands are unaffected in terms of price elasticity. The innovator has, however, a higher elasticity than all other brands; the study provides no explanation for this circumstance. Following the differentiation argument, the innovator being the most differentiated, should have the lowest price elasticity but other factors like the risk of an unknown product also play a role.

5.1.3 Premium Positioning and Quality

Another determinant that can be used to gain insights is the premium positioning of brands. The results of previous research are mixed. The influence might differ especially between a price decrease and a price increase.

At times premium brands are defined in terms of their price. One of the leading market research companies GfK defines a premium brand as higher priced than the market leader (Twardawa 2010), this is in line with the distinction of premium brands vs. low-priced brands by Bemmaor (1984) and Mulhern/Williams/Leone (1998), where no further definition is provided. Another aspect to be considered is that usually the level of advertising is higher for premium brands. Premium brands compete mainly through advertising while economy brands compete mainly on price (Bemmaor 1984; Carpenter et al. 1988).

Some studies assume that price and quality are highly correlated, while in other studies, the price level is not necessarily positively related with quality. For example, economic models are used to objectively measure consumers' perception of quality (Allenby/Rossi 1991). The quality level is often assumed to be higher for premium brands (Blattberg/Wisniewski 1989). For further information on price as a quality indicator and correlations of price and quality it is referred to Diller (2008, p. 150-154), Gavious/Lowengart (2012) and Völckner/Hofmann (2007).

Since there is no clear distinction between quality, price level and premium positioning, these studies are pooled together, as displayed in table (5-3). Even though there are mixed results, the overall picture shows the tendency that premium/higher quality brands have higher price elasticities than lower priced economy brands.

Table 5-3: Determinant – Premium Positioning and Quality

Study	Product Category	Key Results	Impact on Elasticity
Allenby/Rossi 1991	margarine	PE high quality brands > PE low quality brands	+
Bemmaor 1984	frequently purchased branded good	PE premium brands > PE low-priced brands	+
Carpenter et al. 1988	household product	PE premium brands < PE economy brands	-
Mantrala et al. 2006	automotive hard parts	PE best quality > PE better quality > PE good quality	+
Mulhern/Williams/ Leone 1998	liquor	PE premium brands vs. PE other brands: n.s.	n.s.

- : negative impact on magnitude of PE
+ : positive impact on magnitude of PE
n.s.: not significant

Asymmetric pattern of switching between premium and economy brands is well documented (e.g., Blattberg/Wisniewski 1989, Kamakura/Russell 1989). Blattberg/ Wisniewski (1989, p. 308) explain the situation the following way: "When higher-price, higher quality brands price deal, they steal unit sales away from other brands in their own price tier and from brands in the tier below (the moderate and private label brands). However, when lower-price, lower-quality brands deal, they draw sales from their own tier (other moderate and private label brands) and the tier below (generics), but in general do not take significant amount of unit sales away from the tier above (national brands)."

Price reductions for higher quality brands attract more customers than do price reductions for lower quality brands. Allenby/Rossi (1991) confirm this finding with scanner panel data on margarine brands. The high quality brands have very high elasticities compared to the low quality brands. The prices are also highest for the high quality brands in this study. In addition, the authors also mention the influence of the market share, i.e. lower market shares of the high quality brand contribute to the higher price elasticities and vice versa (Allenby/Rossi 1991, p. 197 f.). At the same level of market share, a higher quality brand will have a higher price elasticity than a lower quality brand due to the interaction between income and substitution effect. Both effects work in the same direction for high quality brands while they work in opposite directions for low quality brands (Allenby/Rossi 1991, p. 197).

Bemmaor (1984) finds that price elasticities are larger for premium brands than for low-price brands using data on a frequently purchased branded product. However, advertising and the price of the product also play a role in this study. Premium brands are highly advertised compared to the low-priced brands and on average 35% more expensive. Given a high level of advertising, premium brand buyers are aware of product benefits (Bemmaor 1984) and may use price less as a quality indicator (Klein/Leffler 1981); and they should therefore react more strongly to a price increase. Buyers of low-priced brands could, however, perceive a quality improvement in a price increase and therefore respond less strongly (Bemmaor 1984, p. 303).

In the automotive hard part retail industry the product categories are composed of several subcategories whereby only one subcategory fits a particular make, model and year of a vehicle. In this special case, the customer enters the market infrequently and buys only one particular hard part that is needed as replacement. Each subclass typically contains three variants that are ordered by quality; a "good" entry grade, a "better" mid-level and a "best" premium brand (Mantrala et al. 2006, p. 590). The pricing structure indicates the existence of price-quality tiers, with the price of the "good" variant being lower than that of the "better" variant which is again lower than the price of the "best" variant in all but one subcategories (Mantrala et al. 2006, p. 594). The average price elasticity of the automotive hard parts is -2.22. If the data is analyzed by quality tiers, there is an average price elasticity of -1.26 for the "good" variant, -1.80 for the "better" and -3.60 for the "best" variant (Mantrala et al. 2006, p. 596).

Contrary results are found by Carpenter et al. (1988). Here the premium brands have lower price elasticities than the economy brands. There is a large variation in price elasticities in the data set. The economy brands have the highest price elasticities. Among the premium brands, those with unique physical characteristics, i.e. claiming tangible benefits to the consumer, have the lowest price elasticities. The other premium brands have price elasticities between the economy and tangible benefit brands (Carpenter et al. 1988, p. 404). No significant differences between premium and other brands can be found in a study on liquor brands (Mulhern/Williams/Leone 1998).

5.1.4 Brand Ownership

Even though the brand ownership was not analyzed or mentioned in most of the original studies, the research at hand examined the price elasticities of private labels in comparison to manufacturer brands. The studies that provide insights are listed in table 5-4.

Table 5-4: Determinant – Private Label vs. Brand

Study	Product Category	Key Results	Impact on Elasticity
Chib/Seetharaman/ Strijnev 2004	cola	Private label has lowest PE (proposed model) Private label has PE among premium brands (nested logit model)	- o
Chintagunta/Honore 1996	saltine crackers	Private label has lowest PE	-
Chintagunta/Jain/ Vilcassim 1991	saltine crackers	Private label has second lowest PE (after market leader)	-
Cooper 1988	coffee	PE private labels < PE brands	-
Kadiyali/Chintagunta/ Vilcassim 2000	orange juice	Private label has highest PE	+
Kamakura/Russell 1989	food item	Private label has highest PE	+
Kim 1995	tuna	Private label has highest PE	+
Kim/Rossi 1994	tuna	Private label has highest PE	+
Reibstein/Gatignon 1984	eggs	Private label has PE among other brands	o

- : negative impact on magnitude of PE
+ : positive impact on magnitude of PE
o : no impact observed

The overall picture is mixed. In four out of nine studies, the private label has the highest price elasticity of the analyzed brands. It is consistent with accepted wisdom that the private label has the highest price elasticity (Kadiyali/Chintagunta/Vilcassim 2000, p. 144). Since the private labels have little to offer other than price one might expect them to have high price elasticities. However, in the remaining five studies the private label has the lowest or a rather low price elasticity. The market shares of the private labels with low price elasticities are sometimes high (Chintagunta/Honore 1996; Chintagunta/Jain/Vilcassim 1991) and other times low (Cooper 1988) and thus cannot explain the variation in price elasticity.

For saltine crackers, the market leader Nabisco has the lowest price elasticity in one study followed by the private label with the second lowest price elasticity (Chintagunta/Jain/Vilcassim 1991). In another study of the same market, the private label has the lowest price elasticity followed by the market leader Nabisco (Chintagunta/Honore 1996). In both studies, Nabisco dominates the market with a market share of over 50% followed by the private label with a market share of approximately 30%.

In the coffee market, as studied by Cooper (1988), the private labels have lower price elasticities than the manufacturer brands. A policy of maintaining a high shelf price and generating sales through price promotions is typical for branded products in the coffee market. This leads to high price elasticities for the main brands. Some generate the majority of their sales during price promotions. The private labels compete primarily on price but with an everyday-low-price strategy, they do not generate enough variation in price to reach the price elasticities of the branded products (Cooper 1988, p. 714).

Strong brands like Pepsi and Coca-Cola also show higher price elasticities than the private label in the cola soft drink market (Chib/Seetharaman/Strijnev 2004). Overall, a more prominent brand may serve a broader set of purposes than a private label, and thus the private label may not substitute effectively even when it is on sale (Shocker/Bayus/Kim 2004, p. 32). The most recent meta-analysis (Bijmolt/Van Heerde/Pieters 2005, p. 146) finds no significant effect of brand ownership on price elasticity. Tellis (1988) did not study this determinant in his meta-analysis.

5.1.5 Direction of Price Change

The effect of a price increase compared to a price decrease is not explicitly studied in any of the research articles. Bijmolt/Van Heerde/Pieters (2005, p. 154) identify the need for further research on separate price elasticities for both directions of price changes.

Two articles listed in table 5-5 can be used to gain insights. In a footnote Bucklin/Srinivasan (1991, p. 65) mention that they use a price cut of 20% to calculate the price elasticities. They also examine a price decrease of 10% and a price increase of 10% and compare the results. The conclusion is that the price elasticities change somewhat with the magnitude and direction of the price change, however, no data is reported to examine the changes. The basic pattern of the results is robust though. The correlations of the price elasticities are rather high. For the 18 brands correlations among the 324-element vectors representing each (18 x 18) elasticity matrix were computed. The correlation of a 20% price cut vs. a 10% price cut is highest with 0.988, which represents almost a perfect correlation of 1. For the same price delta of 10% but opposite directions of the price change, the correlation of a 10% price cut vs. a 10% price increase is also rather high with 0.927. Slightly lower is the correlation when both the price delta and the direction of price change are different. Looking at a price cut of 20% vs. a price increase of 10% the correlation is 0.893.

Table 5-5: Determinant – Direction of Price Change

Study	Product Category	Key Results	Impact on Elasticity
Bucklin/Srinivasan 1991	coffee	▪ Comparison of PE with price changes of -20%, -10% and +10%: pattern of results is robust, high correlations ▪ Price elasticities change somewhat, but no data reported	o ?
Moon/Russell/ Duvvuri 2006	toilet tissues	▪ PE 10% price increase > PE 10% price cut	+

+ : positive impact on magnitude of PE
o : no impact observed
? : unknown impact due to lack of data

Moon/Russell/Duvvuri (2006) never state in their study that the price elasticities for the price increase are higher than the price elasticities for the price reduction. However, looking at table 5-6, one can see that the elasticities for the price increase are for almost all cases (87%) higher than for the price decrease. Given the limited data availability, no detailed analysis of significant differences between each brand can be conducted. However, an analysis of the means shows a significantly larger price elasticity for the price increase compared to the decrease for the consumers with no reference price mechanism (two tailed sign. level $p < 0.05$). The tendency is also supported comparing the elasticities of consumers with a memory based reference price (two tailed sign. level $p < 0.10$). Also in this study the pattern of results is robust in the price decrease and price increase scenario (Moon/Russell/Duvvuri 2006, p. 8).

Table 5-6: Determinant – Direction of Price Change – Study Example

Brand	Elasticities: 10% price DECREASE			Elasticities: 10% price INCREASE		
	No RP*	Memory based RP	Stimulus based RP	No RP	Memory based RP	Stimulus based RP
Charmin	-4.77	-4.92	-2.86	-6.04	-5.91	-4.06
Northern	-4.13	-4.84	-3.40	-4.50	-4.92	-3.98
Cottonelle	-2.47	-6.49	-3.67	-5.25	-7.26	-6.04
White Cloud	-4.67	-5.46	-3.99	-5.22	-6.01	-3.58
Family Scott	-3.31	-4.62	-1.44	-4.42	-4.57	-2.19
Average	**-3.87**	**-5.26**	**-3.06**	**-5.09**	**-5.73**	**-3.97**

RP: reference price
Source: adapted from Moon/Russell/Duvvuri 2006

The focus of the study of Moon/Russell/Duvvuri (2006) is on the question how consumers encode price information, via different reference price mechanism, in making a purchase decision. This leads to the next determinant – customer characteristics.

5.1.6 Customer Characteristics

Customer characteristics are a frequently studied subject in price elasticity research. Eight studies look at various aspects of these determinants (table 5-7).

Table 5-7: Determinant – Customer Characteristics

Study	Product Category	Key Results	Impact on Elasticity
Hoch et al. 1995	packaged consumer goods	▪ % of population over 60 years old: mixed effects on PE* ▪ % of population with college education: reduces PE ▪ % of households with 5 or more members: increases PE ▪ log of median income: mixed effects* ▪ % of houses with a value > $150,000: reduces PE ▪ % of women who work: increases PE ▪ % of black or Hispanic consumers: increases PE	+/- - + +/- - + +
Kim/Rossi 1994	tuna	▪ consumers with high purchase frequency: increases PE ▪ consumers with high purchase volume: increases PE	+ +
Krishnamurthi/Raj 1991	coffee	▪ loyalty: reduces choice elasticity ▪ loyalty: increases quantity elasticity ▪ loyalty overall effect: reduces total elasticity	- + -
Mehta/Rajiv/ Srinivasan 2003	detergent	▪ consumer with large consideration set: increases PE	+
Montgomery 1997	orange juice	▪ % of population over 60 years old: increases PE ▪ % of population with college education: increases PE ▪ % of households with 5 or more members: increases PE ▪ log of median income: increases PE (weak effect) ▪ % of houses with a value > $150,000: reduces PE ▪ % of women who work: reduces PE ▪ % of black or Hispanic consumers: increases PE	+ + + + - - +
Moon/Russel/ Duvvuri 2006	toilet tissues	▪ PE memory based reference price consumers > PE no reference price consumers > PE stimulus based reference price consumers	n/a
Mulhern/Williams/ Leone 1998	liquor	▪ markets with higher incomes: increases PE ▪ % of Hispanic consumers: n.s. ▪ % of black consumers: reduces PE	+ n.s. -
Murthi/Srinivasan 1999	ketchup	▪ evaluation of all relevant information on brands: increases PE	+

- : negative impact on magnitude of PE
+ : positive impact on magnitude of PE
* equal number of positive and negative effects
n.s.: not significant
n/a: not applicable

As mentioned in the previous chapter (compare table 5-6), Moon/Russell/Duvvuri (2006) find that the price elasticities of customers depend upon how reference prices are formed and how they are used to make a purchase decision. They distinguish three different reference price mechanisms of consumers. Price elasticities of consumers with a memory based reference price are lager than price elasticities of consumers with no reference price which in turn are larger than the price elasticities of consumers with stimulus based reference prices. This pattern holds true for both directions of price changes (Moon/Russell/Duvvuri 2006, p. 8). This leads to the conclusion that consumers who recall prices from memory because they monitor the pricing environment are the most price sensitive.

Consumers do not evaluate all relevant information on brands on all occasions. Overall in more than 40% of the purchase occasions, customers are unaffected by price, since they do not evaluate this information. The customer segment that is highly likely to evaluate information responds to price three to four times stronger than the low probability segment (Murthi/Srinivasan 1999, p. 233). In the case of non-evaluation consumers rely on brand preferences, past evaluation of brands and loyalty (Murthi/Srinivasan 1999, p. 232). Limited information processing abilities lead to the formation of consideration (or evoked) sets. Consumers with larger consideration sets tend to be more price sensitive. Price elasticities are underestimated when it is assumed that consumers get price information on all brands at zero cost (Mehta/Rajiv/Srinivasan 2003, p. 60).

"Conventional wisdom suggests that consumers who are loyal to a brand will be insensitive to the brand's price" (Krishnamurthi/Raj 1991, p. 172). An empirical analysis of the coffee market and another undisclosed frequently purchased product found a more comprehensive relationship between loyalty and price elasticity (Krishnamurthi/Raj 1991). On the one hand, loyal consumers are less price sensitive in their choice decision. Because of the strong preference for the brand they will usually chose the brand regardless of its price. On the other hand, they are more price sensitive in the quantity decision than non-loyal consumers. They adjust the quantity of purchase depending on the price to take advantage of lower prices. Looking at the overall effect, loyal consumers are less price sensitive than non-loyal consumers. The specific characteristics of the coffee market in terms of high purchase volume during promotion was mentioned in the analysis of Bucklin/Srinivasan (1991) which studied essentially the same brands dominating the coffee market. This should be taken into account when considering generalizations.

Hoch et al. (1995) and Montgomery (1997) studied besides the same competitive variables also identical consumer characteristics. The results are somewhat mixed. Three characteristics have consistent results. In areas with households of more than five family members the price sensitivity is higher. The same is true in areas with a high percentage of black and Hispanic consumers. As the percentage of houses with a value of more than $150,000 increases, price sensitivity decreases. No clear direction was to be found for the characteristics age over 60 years, college education, income and women who work. The combination of customer and competitive characteristics, which however add little over the customer characteristics, explain 22% of the variance for orange juice (Montgomery 1997) and on average 67% of the variance across all products raging from 34% for toothpaste to 86% for frozen dinners (Hoch et al. 1995, p. 24).

Households can be also classified into different life stages according to characteristics like marital status, age, household size, employment, etc. (Du/Kamakura 2006). A study on consumers' budget allocation simulates a 50% price increase for motor fuel as well as gas, heating oil and coal. Results show that households in different life stages and with different income levels vary greatly in demand changes and budget allocation across various product categories due to the price increase. Households with higher incomes are less elastic in their demand for motor fuel than households with relatively low incomes (Du/Kamakura 2008, p. 124 f.). The difference between the poorest and richest quintiles was strongest for the life stage category of widowed elderly people.

The findings of Hoch et al. (1995) and Montgomery (1997) that black and Hispanic consumers are more price sensitive are not supported by Mulhern/Williams/Leone (1998) who find no significant effect on the influence of Hispanic consumers and a lower price sensitivity in areas with a high percentage of black consumers. There is also no clear tendency how income influences price sensitivity [Hoch et al. 1995 (equal number of positive and negative effects), Huang/Jones/Hahn 2007 (PE decrease), Montgomery 1997 (PE increase), Mulhern/Williams/Leone 1998 (PE increase)].

5.2 Insights from Consulting Projects

In order to gain insights from the consulting project data on determinants, in the first step expert interviews were conducted with experienced consultants specialized in price management. A broad array of industry experts were interviewed regarding determinants and availability of information contained in the consulting project data. Based on this information the selection and coding of the determinants were carried out. Following this, the consulting project data was analyzed with regard to the selected determinants. Hereby the data on the average price elasticities as well as the data on the price elasticities for a price increase and the data on the price elasticities for a price decrease were utilized for an assessment of the determinants. In addition, industry specific aspects were assessed.

5.2.1 Selection of Determinants

Based on the findings in the academic data set and the literature review, expert interviews with experienced consultants in the price management field were conducted. 14 expert interviews with partners and directors (i.e. one hierarchy level below the partner level) of the consulting company Simon-Kucher & Partners were conducted. A broad array of industries (automotive, medical technology, pharmaceuticals, telecommunications, FMCG, consumer durables, logistics, financial services, technology, etc.) was covered by the corresponding experts in that field. Some interview partners have worked in the price management field for over 20 years and thus have extensive experience regarding price elasticities.

More than 30 variables covering product, market and customer characteristics were assessed in qualitative interviews regarding their influence on price elasticities. The direction of the effect and the strength of the effect on price elasticities were discussed. In addition, the experts provided information on their industry experience and provided examples based on their project experience to illustrate the influence of specific determinants. In open discussions, other potential influencing factors were raised and project examples were given as a reference. Furthermore, the available information in the project data and thus the feasibility of the coding procedure was discussed.

The interview process led to the decision to focus on product characteristics and market characteristics. Customer characteristics were not considered due to the limited availability of information in the project data and the fact that several factors viewed from the customer perspective here are also reflected in a similar way in the product characteristics. Customers' price elasticities might vary across customer segments or with individual characteristics, e.g. loyal customers might react to a price increase less price sensitively than non-loyal customers. Customers who perceive a higher risk in buying the product might be less price sensitive than customers who perceive the transaction as low risk. The price elasticities were not calculated and not analyzed on an individual customer level or for specific customer segments, e.g. loyal vs. non-loyal customers. The overall price elasticity for a product was assessed; not price elasticities for specific customer groups.

In addition, the customer characteristic perception of risk might be related to the product characteristic complexity of the product; and the price knowledge of the customer might also be related to price transparency, etc. Overall, the pricing experts evaluated the following customer characteristics: brand loyalty, price knowledge, perception of risk, importance of convenience, importance of image and prestige, appreciation of quality, customer satisfaction, perception of price fairness, product knowledge. The results of the assessment can be found in the appendix (cf. Appendix D-1 to Appendix D-6).

Chapter 5: Determinants of Price Elasticity 101

To extend the range beyond previous meta-analytic designs, this research endeavors to incorporate diverse variables in order to explore additional determinants of price elasticities. As described before, the focus of previous research was on research methodology and limited insights could be drawn regarding market and product characteristics. The focus of this research is on the product and market characteristics, while several aspects of the research methodology and data characteristics are also included to control for these effects. This leads to the general framework displayed in figure 5-2.

Figure 5-2: Research Framework for Determinants of Price Elasticity

Product and Market Characteristics
- Degree of Differentiation
- Quality
- Type of Brand
- Absolute Price Level
- Stage of Product Life Cycle
- Level of Competition
- Price Transparency
- Level of Complexity
- Frequency of Purchase
- Type of Market

Research Methodology
- Price Anchor
- Volume Measurement
- SKU vs. Brand
- Method Used

→ Level of PE

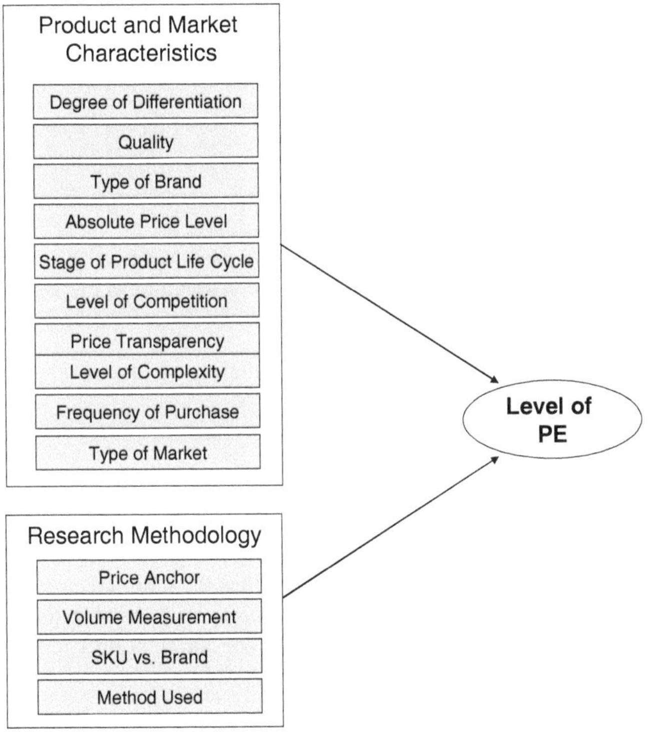

The research at hand concentrates in line with Bijmolt/Van Heerde/Pieters (2005) on the following two categories; the first category comprises product and market characteristics, the second category comprises data characteristics and research methodology. The research scope is broadened especially by adding previously not examined determinants.

The focus of the analysis of determinants is on product and market characteristics. Bijmolt/Van Heerde/Pieters (2005) and the majority of research articles on price elasticities have sufficiently covered aspects of the research methodology, a further analysis would therefore provide limited insights. In addition, there is a restricted ability to derive insights for managerial decision making from these determinants.

It is essential for creating a successful marketing strategy to understand how price elasticities vary with product and market characteristics (Bijmolt/Van Heerde/Pieters 2005, p. 141). New variables are added in order to expand the research scope and gain additional knowledge on their relationship with price elasticities. The new variables are the degree of differentiation, the quality, additional types of brands, the absolute price level of the product, the level of competition, the price transparency, the level of complexity, the frequency of purchase and the type of market. Another variable is the stage of product life cycle which was also studied by Bijmolt/Van Heerde/Pieters (2005). All other variables are new.

In the second category research methodology, the author assesses whether aspects of the research process influence the price elasticity. The research examines the influence of the price anchor used to calculate the price elasticity, the influence of the volume measurement in the price response and the method used to derive the price response function. In addition, the influence of modeling the research object at the SKU vs. the brand level is analyzed in line with Bijmolt/Van Heerde/Pieters (2005).

The variables will be explained in more detail in the coding section which is covered in the next chapter (5.2.2).

5.2.2 Coding of Determinants

A coding sheet was created, which was applied to all price elasticity cases in the consulting project data set. The coding sheet specifies the information to be extracted from each project (Lipsey/Wilson 2001, p. 73). The coding was sometimes discussed with the industry expert who had conducted the consulting project to gain insights, especially for some very specific products and industries.

Coders must have considerable background in the methodology and the specific research domain at issue to perform the coding task well. They must not only understand the coding protocol in detail and depth but must also have the knowledge and skills to properly read and interpret the data reports (Lipsey/Wilson 2001, p. 88). Access to the data was restricted due to confidentiality agreements. In some cases the author did not gain full access to all project data due to confidentiality aspects. In these cases the consultant who conducted the project performed the coding with guidelines provided and the coding was discussed.

Chapter 5: Determinants of Price Elasticity   103

In general, the coding was performed by the author who made herself familiar with the product or service by reading background information on the project, final project reports and having discussions with the project members. Generally, the coder agreement and coder reliability are typically quite good, even for more complex coding tasks (Cooper 1998, pp. 95-97). It is common for smaller studies that the coding is entirely done by the researcher (Lipsey/Wilson 2001, p. 90).

The data was coded using the following determinants with the corresponding scales and levels. As all the price elasticity cases are on the brand level or even stock-keeping unit level, a product category per se was not coded but the coding was performed for a specific product on a more detailed brand or SKU level.

PRODUCT AND MARKET CHARACTERISTICS

Degree of Differentiation

The degree of differentiation is the perceived distinctiveness of the product at the brand level. It defines the brand and reflects its ability to stand out from the competition (Mizik/Jacobson 2008, p. 16). It is defined by how the product offering, as perceived by the consumer, differs from its competition on any physical or non-physical product characteristic (Porter 1985; Dickson/Ginter 1987). Perceptual differences are created by usage experience, word of mouth, image and promotion; actual differences are created by product characteristics (Carpenter/Glazer/Nakamoto 1994; Dickson/Ginter 1987; Shostack 1977). A main question is whether the brand is unique and distinctive (Mizik/Jacobson 2008; Lee/Tang 1997). The degree of differentiation comprises the substitutability of a product. This is essentially a reverse relationship. If the substitutability is very high the degree of differentiation is very low and vice versa. The substitutability is an aspect of the degree of differentiation and indicates how easy is it to replace one product with another, also looking at brand characteristics.

The variable is coded on a 7-point scale ranging from 1 = very low to 7 = very high degree of differentiation.

	very low	medium	very high
degree of differentiation	(1) --- (2) --- (3) --- (4) --- (5) --- (6) --- (7)		

The coding of this research differentiates the positioning of the product based on the type of brand and evaluates the quality of the product in a separate variable. This distinction is often not made and it is assumed that premium brands have a higher quality (cf. chapter 5.1.3)

Quality

Quality is a global assessment about the superiority or excellence of a product, it is evaluated at a higher level of abstraction than a specific attribute of a product (Zeithaml 1988). The overall quality of a product at the brand level is assessed using a 7-point scale (Aaker/Keller 1990; Völckner/Hofmann 2007). The range was kept from 1 = very low to 7 = very high but a more specific internal guideline was created of what type of quality the levels represent in detail.

(1) very low quality
(2) low quality
(3) mass market quality
(4) standard quality
(5) high quality
(6) premium quality
(7) luxury quality

	very low	medium	very high
quality	(1) --- (2) --- (3) --- (4) --- (5) --- (6) --- (7)		

Type of Brand

The research at hand differentiates not only private labels and manufacturer brands as Bijmolt/Van Heerde/Pieters (2005) did, but implements a more detailed coding in the following five types of brands: private label, premium private label, manufacturer brand, premium manufacturer brand, and finally luxury manufacturer brand. The private label and the premium private label were eliminated later on since no cases fitted this category.

	private label	premium private label	manufacturer brand	premium manufacturer brand	luxury manufacturer brand
type of brand					

Absolute Price Level

The absolute price level reflects the price of the product as tested in the primary research. Different measures are used in the original consulting projects to reflect how customers think when processing price information. This was typically tested in pilot interviews and preliminary discussions and then adjusted for country specific needs,

Chapter 5: Determinants of Price Elasticity 105

e.g. price per day, month or quarter, reflecting the decision making parameters of customers.

The majority of prices were in Euro, other currencies were converted to Euro with the average exchange rate of the year the primary research was conducted. First, the actual price of the product was put on record and second, to minimize the spread, the prices were coded in the following seven categories.

(1) 0.00 € – 5.00 €
(2) 5.01 € - 50.00 €
(3) 50.01 € - 100.00 €
(4) 100.01 € - 1,000.00 €
(5) 1,000.01 € - 10,000.00 €
(6) 10.000.01 € - 50,000.00 €
(7) above 50,000.00 €

	very low medium very high
absolute price level	(1) --- (2) --- (3) --- (4) --- (5) --- (6) --- (7)

Stage of Product Life Cycle

The stage of the product life cycle is evaluated at the brand's product line level, e.g. a specific automobile brand's series (Volkswagen Golf, BMW 3 series), not the life cycle of the product category (automobile) in general or the manufacturer's brand in general. Simon (1979) notes that the assessment on the brand vs. product level is essential to capture the price elasticity effects accurately. e.g. the introductory stage of a new brand of laundry detergent vs. the mature life cycle stage of laundry detergent. Simon (1979) and Tellis (1988) name this determinant to be more precise brand life cycle, whereas Bijmolt/van Heerde/Pieters (2005) refer to the determinant as product life cycle and compare it directly to the results of Tellis (1988). The research at hand follows this more frequently used terminology and distinguishes the commonly accepted four stages of the product life cycle: introduction, growth, maturity and decline. For the data analysis, the introduction and growth phase are grouped together as the early stage of the product life cycle and the maturity and decline phase are combined in the late stage as in Tellis (1988) and Bijmolt/Van Heerde/Pieters (2005).

	intro-duction	growth	maturity	decline
stage of product life cycle				

Level of Competition within Market

The level of competition within the market is determined by whether there are only a few brands and one brand dominates the market or whether there is intense competition for market share (Bell/Chiang/Padmanabhan 1999). The level is also assessed based on the number of competitors, whether the market is a monopoly or if there is a moderate level of competition or even price wars.

The intensity of the competition with in the market is measured on a 7 point scale form 1 = very low to 7 = very high level of competition.

	very low	medium	very high
level of competition	(1) --- (2) --- (3) --- (4) --- (5) --- (6) --- (7)		

Price Transparency

Price transparency exists when customers can easily get a clear, comprehensive, current and effortless overview about a product's prices (Matzler/Würtele/Renzl 2006, p. 222; Diller 1997). This implies low search cost to get information on prices. A complex pricing structure lowers price transparency, whereas a simpler pricing structure increases price transparency. The increased usage of the internet and price comparison portals facilitates price transparency (Simon/Fassnacht 2009, p. 520).

The price transparency is measured on a 7 point scale form 1 = very low to 7 = very high price transparency.

	very low	medium	very high
price transparency	(1) --- (2) --- (3) --- (4) --- (5) --- (6) --- (7)		

Level of Product Complexity

The product complexity is defined as the extent to which the consumer perceives a product to be difficult to understand or use (Rogers 1995). The perceived complexity of a product is related to the number of product attributes (Swaminathan 2003, p. 95) and the amount of information that must be gathered to make an accurately evaluate the product (McQuiston 1989, p. 70).

A common way to enhance a product is by increasing its number of features (Goldenberg et al. 2003; Mukherjee/Hoyer 2001; Nowlis/Simonson 1996), which

Chapter 5: Determinants of Price Elasticity 107

provides greater functionality for consumers. A product that offers a large number of options or that involves a large number of steps in its use will typically be seen as more complex (Burnham/Frels/Mahajan 2003, p. 113).

The product complexity is coded on a 7-point scale ranging from 1 = very low to 7 = very high product complexity.

	very low	medium	very high
product complexity	(1) --- (2) --- (3) --- (4) --- (5) --- (6) --- (7)		

Frequency of Purchase

In the literature there are many ways to define and operationalize frequency of purchase, for example as the number of purchase occasions divided by the length of time, e.g. in weeks (Kim/Rossi 1994), or the average number of days between consecutive purchases of the category (Ailawadi/Harlam 2004, p. 151). Other options to operationalize the frequency of purchase is to measure the average number of purchases made by purchasers in a specific category in a one-year period (Estelami/Lehmann/Holden 2001, p. 347) or rate the frequency of purchase on a 5-point scale from 1 = never to 5 = always (Schlegelmilch/Bohlen/Diamantopoulos 1996).

The frequency of purchase is coded according to the following categories on a 7-point scale. The range from 1 = very low to 7 = very high was kept, but in addition a more specific internal guideline of what time frame the levels represent in detail was created.

(1) more than 10 years between purchases
(2) every 6 to 10 years
(3) every 2 to 5 years
(4) yearly
(5) every few months
(6) every few weeks
(7) every few days

	very low	medium	very high
frequency of purchase	(1) --- (2) --- (3) --- (4) --- (5) --- (6) --- (7)		

Type of Market

The type of market is coded into business-to-business and business-to-consumer market as the buying behavior differs distinctively in these two markets (Johnston/Lewin 1996, Webster/Wind 1972).

	business-to-business	business-to-consumer
type of market		

RESEARCH METHODOLOGY

Price Anchor

The price anchor is the price used as the starting point for both the price increase and the price decrease for the price elasticity calculation. The calculation procedure used the current price of the product as the price anchor whenever possible. In general, it is most often used when the product is already on the market. If the product is not on the market, alternative price anchors have to be used. If an antecessor product existed on the market, then the price of this product is usually used, especially for next-generation products. In other cases when no obvious anchor can be identified the profit-optimal price is used as the reference point on the price response function. Another option for a price anchor is the price used in the base case scenario, when several price points and scenarios are evaluated. This is usually the price at which the manufacturing company plans to introduce the product.

	current price	price of antecessor	profit optimal price	base case price
price anchor				

Volume Measurement

The volume measurement refers to how sales are measured in the price response function and thus the underlying volume for each price point used to calculate the price elasticity. Essentially, the volume is either measured in absolute sales like number of sold units or relative sales like market share. However, in some projects the initial volume that corresponds with the price anchor, e.g. the current price, is set to an index of 100. This can represent an absolute sales index or a market share index and then it is evaluated how this index changes with the price increases or price reductions. The variables are grouped into two categories, absolute volume measurement (sales) and relative volume measurement (market shares, indices).

Chapter 5: Determinants of Price Elasticity 109

	absolute sales	absolute sales index	relative sales	relative sales index
volume measurement				

Item Level

It is coded whether the research item is defined at the brand level or stock-keeping unit level.

	SKU	brand
item level		

Method Used to Derive Price Response Function

This refers to the method that was used to derive the price response function in the primary research. Direct methods include procedures like expert interviews with the PRICESTRAT tool and direct questioning of volume bought or expected sales at different price points. Indirect methods are primarily conjoint and discrete choice models.

	direct	indirect
method used		

General Information about the Project

In addition, other information about the projects is also put on record. For example, information on the product in general (e.g. yogurt) and also with specific details (e.g. Dannon yogurt 12 oz, strawberry flavor), the industry, the project team, the month and year the project took place.

5.2.3 Expected Relationship of Determinants with Price Elasticity

The previous literature focuses on short-term price elasticities and not long-term price changes of the regular price. Therefore, strictly speaking, the results of previous research cannot be relied upon (Bijmolt/Van Heerde/Pieters 2005; Fok et al. 2006, p. 455). However, previous literature is drawn on to provide some guidelines for variables assessing long-term regular price changes. In addition, new variables are explored. This is a procedure commonly used in meta-analyses updating previous research (Bijmolt/Van Heerde/Pieters 2005; Sethuraman/Tellis/Briesch 2011).

Bijmolt/Van Heerde/Pieters (2005) did not explicitly formulate hypotheses about the expected relationship of determinants and price elasticity. In addition, no reasoning is provided for the inclusion of additional determinants being absent at Tellis' (1988) study (Bijmolt/Van Heerde/Pieters 2005, p. 141 f.). A review of recent meta-analyses for personal selling elasticities (Albers/Mantrala/Sridhar 2010) and advertising elasticities (Sethuraman/Tellis/Briesch 2011) also shows that it is common to have no prior expectations of determinants included in the model. Sethuraman/Tellis/Briesch (2011, p. 459) state that "availability of new data permits us to investigate several new variables" but also provide no further reasoning on the selection of variables. They collected data on variables that "could potentially influence elasticity" but they explicitly state that they have "no prior expectations" for several variables and that they are "unable to predict the sign of correlation" for other variables (Sethuraman/Tellis/Briesch, p. 460 f.). Also Albers/Mantrala/Sridhar (2010, p. 843 f.) have for more than half of their included variables no hypotheses regarding the effect on personal selling elasticity. Therefore, the research at hand follows commonly accepted practice in meta-analytic research on elasticities.

PRODUCT AND MARKET CHARACTERISTICS

The first category of determinants comprises product and market characteristics. Besides previously researched aspects, the research is broadened to examine new variables, such as the degree of differentiation and the level of complexity.

Degree of Differentiation

The degree of differentiation indicates whether a product offering is able to differentiate itself from other offerings based on physical and non-physical product characteristics; it indicates whether it is unique and distinct (Dickson/Ginter 1987; Lee/Tang 1997; Mizik/Jacobson 2008; Porter 1985).

As the perceived distinctiveness of the product and the ability to stand out from the competition increases, the price elasticity of consumers is expected to decrease. The substitutability is low and therefore it is hard to replace the product with another offering, supporting the expectation that the price elasticity will be lower. Research supports this expectation. A low degree of differentiation leads to higher price sensitivities (Carpenter et al. 1988; Ramirez/Goldsmith 2009; Van Heerde/Mela/ Manchanda 2004); whereas a higher degree of differentiation leads to lower price elasticities (Allenby 1989; Mitra/Lynch 1995). In the expert interviews this relationship between degree of differentiation and price elasticity had the highest level of agreement and the strongest effect expected, thus supporting the expectation.

Expected relationship with price elasticity:
The degree of differentiation is expected to lower the price elasticity.

Quality

The role of the quality is still not clear in pricing research. There are many conflicting results and it is hard to differentiate the quality aspect from the premium branding aspect. In this research, the branding and the quality of the product are treated separately to get a deeper understanding of the influencing factors.

Intuitively, it is assumed that products with a higher quality level have a relatively low price elasticity, which is, however, not necessarily true (Simon/Fassnacht 2009, p. 109). It is, for example, well documented that higher quality products tend to attract more customers than lower quality products when there is a price deal (Allenby/Rossi 1991; Blattberg/Wisniewski 1989; Kamakura/Russell (1989).

Mantrala et al. (2006) also find higher price elasticities for higher quality products. Quality tiers were created and the average price elasticities for the "good" quality were lowest, the "better" quality significantly higher and the "best" quality displayed an average price elasticity that was twice as high as the "better" quality tier looking at the magnitude of price elasticity. The research at hand examines not temporary price deals but long-term price changes of the regular price. However, it is possible that higher quality products attract a broader customer base than lower quality products. On the other hand, Lynch/Ariely (2000) show that having more information on the quality of a product can reduce the price elasticity.

Expected relationship with price elasticity:
No prior expectation

Type of Brand

The research at hand differentiates not only between private labels and manufacturer brands as Bijmolt/Van Heerde/Pieters (2005), but has a more detailed differentiation. Previous research examined private labels and premium brands vs. manufacturing brands. Since this research contains no data on private labels, the focus is on manufacturing brands vs. premium and luxury brands. For luxury brands, the price elasticities are not well researched yet and the pricing procedure displays unique characteristics (Simon/Fassnacht 2009, p. 64-67).

Research results are ambiguous where the influence of the type of brand is concerned. Premium brands have lower price elasticities (Carpenter et al. 1988), premium brands have higher price elasticities (Bemmaor 1984) and premium brands do not show significant differences to other brands (Mulhern/Williams/Leone 1998). Bijmolt/Van Heerde/Pieters (2005) found no significant differences between manufacturer brands and private labels.

Expected relationship with price elasticity:
No prior expectation

Absolute Price Level

The absolute price level reflects the price of the product as tested in the primary research. Bijmolt/Van Heerde/Pieters (2005, p. 146) find higher levels of price elasticities for items with higher price levels such as consumer durables than for groceries. However, in the meta-analysis of Tellis (1988) detergents have higher price elasticities than durables.

Bijmolt/Van Heerde/Pieters (2005, p. 146) examine groceries vs. consumer durables and state that their finding "is consistent with the notion that consumers have a stronger incentive to respond to price changes of big-ticket items (durables) than for smaller ticket items (groceries)". Expert interviews displayed a substantial variation in responses. 33% of the experts disagreed with the notion that products with a higher absolute price, i.e. big ticket items, display a larger price elasticity, while 42% of respondents agreed and the last 25% of respondents would neither agree nor disagree with this statement. Arguments went in both directions. For example, some argued that if you have a high investment then every cent counts, while others suggested that not the absolute price level determines the price elasticity but the type of product, e.g. is it a car vs. a toothpaste a customer buys or is it a computer tomography vs. standard wound care a hospital has to decide for.

Expected relationship with price elasticity:
No prior expectation

Stage of Product Life Cycle

Previous meta-analyses provide conflicting results regarding the effect on price elasticity. Bijmolt/Van Heerde/Pieters (2005) find lower price elasticities in the maturity and decline phase, whereas Tellis (1988) finds lower price elasticities in the introduction and growth phase of the product life cycle.

Overall, products display rather different price elasticity patterns over their life cycle, some decrease, some increase, others increase first and then decrease in the later stages or the other way round (Parker/Neelamegham 1997). It might also depend on the initial price setting and the pricing strategy for a product, e.g. skimming or penetration, whether price elasticities are increasing or decreasing over time (Monroe 2003, p. 369). The interviews with pricing experts supported the assumption of various life cycle patterns.

Expected relationship with price elasticity:
No prior expectation

Level of Competition within Market

Previous analyses of the level of competition (Hoch et al. 1995; Montgomery 1997; Mantrala et al. 2006; Van Heerde/Mela/Manchanda 2004) yielded mixed results regarding the influence on the level of price elasticity. The competitive characteristics added little explanatory power to the magnitude of price elasticity, the determinants yielded weak and mixed effects, i.e. the same determinant on competition lowered price elasticity in some product categories and increased price elasticity in other product categories (Montgomery 1997, p. 321; Hoch et al. 1995, p. 24).

Higher levels of competition with price wars can lead to increased price sensitivity of consumers to weekly store prices, even though price elasticities are not explicitly assessed, there is support for the assumption that competition can lead to higher price elasticities (Heil/Helsen 2001; Van Heerde/Gijsbrechts/Pauwels 2008). It also depends on how smart the market is. When one company raises the prices, do the others follow (smart) or do they lower their prices to gain market share (not smart). Intuitively, higher level of competition lead to higher price elasticities; this is supported by the view of pricing experts.

Expected relationship with price elasticity:
The level of competition is expected to increase the price elasticity.

Price Transparency

Price transparency exists when customers can easily get a clear, comprehensive, current and effortless overview about a product's prices (Matzler/Würtele/Renzl 2006, p. 222; Diller 1997).

A consequence of high price transparency is that customers have lower search and evaluation costs (Matzler/Würtele/Renzl 2006, p. 219). Price transparency has increased due to internet shopping agents and price comparison portals that provide information on prices and stores (Simon/Fassnacht 2009, p. 520). These intermediaries reduce consumers' search cost more than 30-fold compared to store visits and telephone inquiries (Brynjolfsson/Smith 2000).

It is commonly assumed that price transparency increases price sensitivity (Simon/Fassnacht 2009, pp. 38, 109). This view is supported by the pricing expert interviews. The rationale is that with low levels of price transparency, price changes are hidden and consumers are less likely to realize the price change, also in more complex tariffs, customers are less likely to reevaluate their choice and are therefore less price sensitive. Lynch/Ariely (2000) find that higher levels of price transparency increase the price elasticity for products commonly available across various online stores. However, price elasticities are not necessarily higher for online than offline purchases. Consumers displayed lower price elasticities for grocery items online than offline (Chu/Chintagunta/Cebollada 2008, p. 290)

Expected relationship with price elasticity:
The price transparency is expected to increase the price elasticity.

Level of Product Complexity

Highly complex products are for the consumer more difficult to understand and harder to evaluate (Mc Quiston 1989, p. 70; Rogers 1995). A larger number of product features and attributes contributes to the level of complexity (Swaminathan 2003, p. 95). In the purchase process of buying more complex products, price is expected to play a less important role and consumers are expected to react less strongly to price changes.

Given a rather complex product, the customer does not understand the product well. Thus even if prices are lowered, the demand is not expected to increase.

Expected relationship with price elasticity:
The product complexity is expected to lower the price elasticity.

Frequency of Purchase

It is assumed in the literature that a high frequency of purchase increases price elasticity (Simon/Fassnacht 2009, p. 108). Consumers with a high frequency of purchase are more price sensitive than consumers with a low frequency of purchase (Kim/Rossi 1994). This could be due to the fact that with higher frequency of purchase, the knowledge of the consumers is better. Whereas for products purchased infrequently there are fewer opportunities for consumers to compare and learn about prices (Ailawadi/Harlam 2004, p. 153).

The research at hand looks at product characteristics and aims therefore to evaluate if products that are bought more frequently have inherently higher price elasticities. Only 50% of experts agree with this statement and the expected strength of the influence is rather low. The other half suggests that this determinant might be more industry specific, e.g. relevant for daily groceries but not when buying a car. An additional argument against the statement above is that in some industries, like telecommunications, contracts are closed for fixed time frames, e.g. two years.

Expected relationship with price elasticity:
No prior expectation

Type of Market

The type of market is distinguished into business-to-business and business-to-consumer market as organizational buying behavior differs distinctively from consumer buying behavior (Johnston/Lewin 1996). Organizational buying behavior is a complex process, involving many people and multiple goals (Webster/Wind 1972, p. 13 f.). Prices are usually not fixed but the result of a negotiation process (Simon/Fassnacht 2009, p. 487).

Purchasing decisions in the business-to-business market are typically conducted by buying centers leading to more complex decision processes and interactions among people compared to the business-to-consumer market. The interpersonal relations between buyers and sellers, negotiations about contracting, specifications of purchasing and tailoring towards individual needs are expected to lead to lower price elasticities in the business-to-business market compared to the business-to-consumer market. Previous research on price elasticities focuses on the business-to-consumer market (Bijmolt/Van Heerde/Pieters 2005)

Expected relationship with price elasticity:
Lower price elasticities are expected for business-to-business products than for business-to-consumer products.

RESEARCH METHODOLOGY

In this part of the research, it is examined whether decisions in the research process influence the price elasticity. This is in line with the procedure of previous meta-analyses (Bijmolt/Van Heerde/Pieters 2005; Tellis 1988). New potentially influencing factors, such as the price anchor or the methodology used, are assessed.

Price Anchor

It is assessed whether the type of price anchor used to calculate the price elasticity has an influence on the price elasticity. In general, the current price is used as the starting point to calculate price elasticities. However, if that is not available, other price anchors such as the price of an antecessor product, the profit-optimal price or the base of a base case scenario are used.

Expected relationship with price elasticity:
No prior expectations

Volume Measurement

In this research, absolute and relative volume measures are assessed. Respondents in the primary research are asked to assess the volume either in absolute terms, i.e. sales, or relative terms, i.e. market share or indices. Price elasticities based on absolute and relative sales, especially market shares, are identical if the price change does not induce primary demand effects. For example, if demand is stimulated through the price change, absolute sales might increase, the market share might be identical since the market expanded but the relative sales, i.e. the market share, of the product stayed the same. In this case the price elasticity based on relative sales would be smaller than that based on absolute sales, as supported by Bijmolt/Van Heerde/Pieters (2005) and Bell/Chiang/Padmanabhan (1999).

However, the data of this research also includes indices of absolute sales, i.e. the current sales volume is set to a baseline of 100 and relative changes to that baseline are assessed; this is a relative measure but based on absolute sales not market share.

In contrast to a purely mathematical calculation, in the primary research respondents are asked to estimate a price reaction and assess volume changes either on relative terms or absolute terms. It might make a difference whether the volume change is assessed on a relative vs. an absolute term. No significant differences are expected. The relationship is tested, however, to gain additional insights.

Expected relationship with price elasticity:
No prior expectations

Item Level

When modeling price elasticities the research item can be assessed at the brand level or the SKU level. In response to a price change a consumer may switch between SKUs. If the analysis is assessed at the brand level, these switches will not be observed. Therefore, the price elasticities assessed at the brand level are expected to be

Chapter 5: Determinants of Price Elasticity 117

lower than those assessed at the SKU level, this is in line with previous research (Bijmolt/Van Heerde/Pieters 2005; Christen et al. 1997). Similar to the fact that the category price elasticity is lower than the brand level elasticity since switches within the product category are not observed (Hoch et al. 1995), the brand elasticity is expected to be lower than the SKU elasticity since switches between SKUs from the same brand are not observed.

Expected relationship with price elasticity:
Lower price elasticities are expected for price elasticities assessed on the brand level compared to price elasticities assessed on the SKU level.

Method Used to Derive Price Response Function

The price response function was derived through primary research using direct methods (such as direct questioning) and indirect methods (such as conjoint measurement). When direct methods are used to derive a price response function, the attention is drawn more to the price compared to indirect methods where price is presented (Simon/Fassnacht 2009, p. 115 f.)

One concern regarding direct methods is that consumers could overestimate price sensitivities since consumers try to give rational answers and could state that they buy more at lower prices and less at higher prices, reflecting a traditional demand curve even though that might not necessarily be the case (Monroe 2003, pp. 222-224). A gap between the verbal statement and the actual behavior is possible (Völckner 2006a). In the willingness-to-pay research indirect and direct methods have been compared in more detail but results also do not clarify which methods leads to higher or lower estimates (Miller et al. 2011; Silva et al. 2007; Voelckner 2006b; Veisten 2007).

Since the research at hand compares two hypothetical approaches, they tend to be more similar (Miller et al. 2011). With indirect methods the relative perceived value of product attributes including price is measured, therefore it is a trade-off analysis. In the data set for this research, the direct interview methodology is not only used with consumers but also industry experts, decision makers und product experts within in the company offering the product or service. Therefore, there are no prior expectations regarding the relationship of the methodology used and the price elasticity.

Expected relationship with price elasticity:
No prior expectations

5.2.4 Analysis of Determinants

In the analysis of the consulting project data, the procedure for the determinants of price elasticities is analogous to that of the magnitude of price elasticities. The determinants are also examined first on an overall basis looking at the average price elasticity and then examined for elasticities based on a price decrease and a price increase separately. Following this, industry-specific aspects are discussed.

First of all, the framework of determinants is analyzed using multiple linear regressions (cf. for this methodology Backhaus et al. 2006, pp. 45-117). "In interpreting these coefficients, it is important to remember that the dependent variable is a negative number. A positive coefficient indicates that as the associated independent variable increases, the price elasticity moves closer to zero" (Hoch et al. 1995, p. 24). Since the terminology of the research at hand looks at the magnitude of price elasticity, a positive coefficient will reduce the price elasticity and a negative coefficient will increase the price elasticity.

Dummy coding is used for the determinants with a nominal scale (cf. chapter 5.2.2). To facilitate the interpretation, the omitted variable that serves as a reference point is listed in the overview. In line with previous meta-analyses (Tellis 1988; Bijmolt/Van Heerde/Pieters 2005) some dummy variables are grouped together, for example, the stages of product life cycle, where the introduction and growth phase represent the early stages and maturity and decline are grouped together representing the late stages of the product life cycle. Details on the coding and grouping of the variable were described previously. Table 5-8 gives an overview of the determinants grouped by product and market characteristic and research methodology. It is indicated when a 7-point scale is used and when dummy coding is used. In addition, for the dummy variable it is indicated which category served as the bases for the interpretation. This dummy variable is dropped and not directly included in the regression analysis. For example, for the type of brand the manufacturer brand serves as the base level and the dummy variable premium brand and luxury brand are included in the regression analysis, so the coefficients will then be interpreted against the base level that serves as a reference.

Chapter 5: Determinants of Price Elasticity 119

Table 5-8: Overview of Variables for Regression Analyses

	Determinants	Coding of Determinants		
Product and Market Characteristics	Degree of Differentiation	7 point scale: 1= very low to 7 = very high		
	Quality	7 point scale: 1= very low to 7 = very high		
	Type of Brand	Manufacturer Brand*	Premium Brand	Luxury Brand
	Absolute Price Level	7 point scale: 1= very low to 7 = very high		
	Stage of Product Life Cycle	Introduction/Growth		Maturity/Decline*
	Level of Competition	7 point scale: 1= very low to 7 = very high		
	Price Transparency	7 point scale: 1= very low to 7 = very high		
	Level of Complexity	7 point scale: 1= very low to 7 = very high		
	Frequency of Purchase	7 point scale: 1= very low to 7 = very high		
	Type of Market	Business-to-Business		Business-to-Consumer*
Research Methodology	Price Anchor	Current Price*	Price of Antecessor	Profit Optimal Price	Base Case Price
	Volume Measurement	Absolute Volume		Relative Volume*
	Item Level	SKU*		Brand
	Method Used	Direct		Indirect*

*indicates dropped dummy variable in regression analyses that serves as a reference

The results of the multiple regression analyses will be shown in the following chapters. For all three analyses – using the average price elasticity, the price elasticity for a price decrease and the price elasticity for a price increase – collinearity was analyzed, assessing the correlations between the determinants as well as the tolerance and variance inflation factors. The analyses did not indicate an issue with multicollinearity.

5.2.4.1 Analysis of Determinants - Average Price Elasticity

First the determinants are analyzed using the average price elasticity. As described before, it is the average of the price elasticity for a price decrease and the price elasticity for a price increase. The average price elasticity served as the basis for the magnitude of price elasticity analysis and is now the foundation of the analysis of determinants. The results of the regression analysis are shown in figure 5-3.

Figure 5-3: Results of Regression Analysis – Determinants Analyzed with Average Price Elasticity

	Determinants	Levels	Standardized β Coefficients
Product and Market Characteristics	Degree of Differentiation	7-point scale	0.335***
	Quality	7-point scale	-0.442***
	Type of Brand	Manufacturer Brand (b)	
		Premium Brand	0.176**
		Luxury Brand	0.237***
	Absolute Price Level	7-point scale	-0.182*
	Stage of Product Life Cycle	Maturity/Decline (b)	
		Introduction/Growth	-0.035 (n.s.)
	Level of Competition	7-point scale	0.082 (n.s.)
	Price Transparency	7-point scale	0.014 (n.s.)
	Level of Complexity	7-point scale	0.206***
	Frequency of Purchase	7-point scale	-0.038 (n.s.)
	Type of Market	Business-to-Consumer (b)	
		Business-to-Business	0.107*
Research Methodology	Price Anchor	Current Price (b)	
		Antecessor Price	-0.009 (n.s.)
		Profit Optimal Price	0.148**
		Base Case Price	-0.050 (n.s.)
	Volume Measurement	Relative Volume (b)	
		Absolute Volume	0.097*
	Item Level	SKU (b)	
		Brand	0.186***
	Method Used	Indirect Method (b)	
		Direct Method	0.193***

$r^2 = 0.270$, adjusted $r^2 = 0.234$, F = 7.60; degrees of freedom = 17, 350; p = 0.000

n = 368
* p< 0.10, ** p< 0.05, *** p< 0.01
Dummy category: type of brand***, price anchor**
(b) indicates base for dummy variables

Chapter 5: Determinants of Price Elasticity   121

The explained variance is 27.0%, this is in the range of previous meta-analyses; Bijmolt/Van Heerde/Pieters (2005) could explain 16% and Tellis (1988) 29%. Also other studies, that do not use a meta-analytic approach, are able to explain a moderate amount of variance, for example 23% (Mulhern/Williams/Leone 1998) and 26% (15% adjusted) (Bolton 1989b). The model fit is satisfactory and highly significant (F = 7.60; degrees of freedom = 17, 350; p = 0.000, n = 368).

Looking at the first determinant group – product and market characteristics – it becomes apparent that the degree of differentiation lowers the price elasticity. This is in line with previous research that indicates that elasticity represents a measure of differentiation (Boulding/Lee/Staelin 1994). The role of non-price advertising in differentiating the product and therefore decreasing customers' price sensitivity is stressed in various articles (Ailawadi/Lehmann/Neslin 2001; Kalra/Goodstein 1998; Kaul/Wittink 1995). Differences in packaging can also be a successful way of differentiating a product, leading to lower price elasticities (Allenby 1989). The market entrance of a new innovative brand decreases the brand differentiation of the existing brand and thus increases the price elasticities of these brands in such a situation (Van Heerde/Mela/Manchanda 2004). Parker/Gatignon (1996) also demonstrate that as a brand becomes less unique over time due to the market entrance of other brands, the price elasticities increase for each brand, whereas price elasticities are initially low for the pioneering brand, then higher for the brands that immediately follow and lower for the later entrants. When customers perceive a low degree of differentiation, the price sensitivity increases (Ramirez/Goldsmith 2009). Products with unique physical characteristics, that provide tangible benefits to the consumer, have lower price elasticities (Carpenter et al. 1988, p. 404).

The positioning based on quality increases the price elasticity, i.e. higher quality leads to higher price elasticities. The fact that higher quality product attract more customers than lower quality products when they price deal, has been documented in the literature (Allenby/Rossi 1991; Blattberg/Wisniewski 1989; Kamakura/Russell 1989). In the current research temporary price changes are not analyzed but a price reduction of a higher quality product might attract a broader customer segment and therefore lead to higher price elasticities than when a lower quality product reduces its price. The quality effect is stronger for the price decrease than the price increase but still holds true for both directions (cf. figure 5-4; figure 5-5). Buyers of low-priced brands could perceive a quality improvement in a price increase and therefore respond less strongly (Bemmaor 1984, p. 303). At a high level of quality buyers are more aware of product benefits and may use price less as a quality indicator (Bemmaor 1984; Klein/Leffler 1981), thus responding stronger to a price increase. The data set contains only limited data on the extreme ends of the quality spectrum. The price elasticity cases concentrate around the following quality levels: standard, high and premium quality (level 4-6 on the 7-point scale).

The dummy category type of brand is significant. Premium brands have significantly lower price elasticities than manufacturer brands. This also holds true for luxury brands compared to manufacturer brands. Private labels and premium private labels are not in the data set and therefore cannot be assessed. Bijmolt/Van Heerde/Pieters (2005) find no significant differences between private labels and manufacturing brands.

The level of complexity also has a significant influence on the price elasticity; with higher levels of complexity the price elasticity is lower. Highly complex products are more difficult for the consumer to understand and harder to evaluate (Mc Quiston 1989, p. 70; Rogers 1995). A larger number of product features and attributes contributes to the level of complexity (Swaminathan 2003, p. 95). In the purchase process of buying more complex products, price plays a less important role and consumers do not react that strongly to price changes. This holds true for price increases and price reductions. Given a rather complex product, the customer does not understand the product well, so even if prices are lowered, the demand does not increase. This view is supported by project examples for complex products with a rather low price elasticity.

The higher the absolute price level, the higher the price elasticity. Consumers become more price sensitive with big ticket items. This is in line with the finding of Bijmolt/Van Heerde/Pieters (2005, p. 146), who examined groceries vs. consumer durables, and state that their finding "is consistent with the notion that consumers have a stronger incentive to respond to price changes of big-ticket items (durables) than for smaller ticket items (groceries)".

Business-to-business products also have lower price elasticities than business-to-consumer products. Organizational buying behavior differs distinctively from consumer buying behavior (Johnston/Lewin 1996). Organizational buying behavior is a complex process and involves many people and multiple goals (Webster/Wind 1972, p. 13 f.). Purchasing decisions in the business-to-business market are typically conducted by buying centers leading to a more complex decision process and interactions among people compared to business-to-consumer market. The interpersonal relations between buyers and sellers, negotiations about contracting, specifications of purchasing and tailoring towards individual needs might be factors contributing to the lower price sensitivities.

No significant influence could be found for the stage of product life cycle, where the previous meta-analyses lead to conflicting results regarding the influence on the price elasticity (Bijmolt/Van Heerde/Pieters 2005; Tellis 1988). Overall, it is hard to generalize because products have rather different price elasticity patterns over their life cycle (Parker/Neelamegham 1997) and the initial price setting and the pricing strategy

Chapter 5: Determinants of Price Elasticity 123

are expected to influence the dynamic of price elasticities over the product life cycle (Monroe 2003, p. 369).

No significant influence could be found for the level of competition, the level of price transparency and the frequency of purchase. These determinants could be explored further in industry analyses to see if these variables gain influence in specific situations. In the overall database, the effects can be leveled out. Previous analyses of the level of competition (Hoch et al. 1995; Montgomery 1997; Mantrala et al. 2006; Van Heerde/Mela/Manchanda 2004) yielded mixed results regarding the influence on the level of price elasticity (cf. chapter 5.2.2). The competitive characteristics added little explanatory power to the magnitude of price elasticity, the determinants yielded weak and mixed effects, i.e. the same determinant (e.g. average distance to the nearest 5 supermarket competitors) lowered the price elasticity in some grocery categories and increased the price elasticity in other grocery categories tested (Montgomery 1997, p. 321; Hoch et al. 1995, p. 24). It also depends on how smart the market is. When one company raises the prices, do the others follow (smart) or do they lower their prices to gain market share (not smart).

The level of price transparency has no significant effect on the price elasticity. One might assume that with a low level of price transparency, price changes are hidden and the customers do not notice the change. Also in more complex tariffs, customers are less likely to reevaluate their choice and are therefore less price sensitive. However, if the prices are non-transparent, customers might also assume that prices must be even lower than the actual price, e.g. in business-to-business logistic services, all clients need to negotiate and there is no official price.

The frequency of purchase does not have a significant effect on the price elasticity. As expert interviews revealed it might be dependent on the type of product. The frequency of purchase could be more relevant in the grocery market, when customers know the prices of everyday products, but in the automotive industry or the business-to-business market, that effect might not be there.

The second determinant group – research methodology comprises type of price anchor used, volume measurement, definition of the research item and method used to derive the price response function. If the profit-optimal price is used as the price anchor compared to the current price as the price anchor, the price elasticity is lower. No significant differences are found for the antecessor price and the base case price. Customers do not react that strongly around the profit-optimal price vs. other points of the price response function. For example, if the product is already overpriced with regard to the perceived value it offers, then a small change in price increase can have a large volume impact. As illustrated in a project example in the away from home food market (figure 4-22), several rather high price elasticities were the result of price anchors used that were far away from the profit-optimal price and lowering the price lead to a substantial increase in demand.

Using absolute volume measurement generates lower price elasticities compared to relative volume measurement including indices; though the influence is rather weak. Price elasticities based on absolute and relative sales, especially market shares, are identical if the price change does not induce primary demand effects. The data of this research, however, also includes indices of absolute sales.

The price elasticity is lower for the brand than the SKU, which is in line with previous research (Bijmolt/Van Heerde/Pieters 2005). Similar to the fact that the category price elasticity is lower than the brand level elasticity since switches within the product category are not observed (Hoch et al. 1995), the brand elasticity is lower than the SKU elasticity since switches between SKUs from the same brand are not observed.

When direct methods are used to derive a price response function, the price elasticity is lower compared to when indirect methods are used. This implies that if consumers or experts are asked directly about the price behavior, the observed price sensitivity is lower than if indirect methods such as trade-offs in conjoint analyses are used to assess the price behavior. This finding is contrary to the belief that consumers tend to overstate their price sensitivity when asked directly. In the data set the direct interview methodology is not only used with consumers but also used with industry experts and key decision makers or product experts within the company offering the product or service.

5.2.4.2 Analysis of Determinants - Price Elasticity for a Price Decrease

After the determinants were analyzed for the average price elasticity, they were also assessed for the opposite directions of the price change, a price decrease and a price increase. First, the price elasticities for a price decrease are utilized as the dependent variable in the regression analysis. The determinants and the standardized coefficients for the price decrease are shown in figure 5-4.

Chapter 5: Determinants of Price Elasticity 125

Figure 5-4: Results of Regression Analysis –
Determinants Analyzed with Price Elasticities for a Price Decrease of 10%

	Determinants	Levels	Standardized ß Coefficients
Product and Market Characteristics	Degree of Differentiation	7-point scale	0.274***
	Quality	7-point scale	-0.403***
	Type of Brand	Manufacturer Brand (b)	
		Premium Brand	0.183**
		Luxury Brand	0.186***
	Absolute Price Level	7-point scale	-0.128 (n.s.)
	Stage of Product Life Cycle	Maturity/Decline (b)	
		Introduction/Growth	-0.043 (n.s.)
	Level of Competition	7-point scale	0.112**
	Price Transparency	7-point scale	0.013 (n.s.)
	Level of Complexity	7-point scale	0.153**
	Frequency of Purchase	7-point scale	-0.064 (n.s.)
	Type of Market	Business-to-Consumer (b)	
		Business-to-Business	0.115*
Research Methodology	Price Anchor	Current Price (b)	
		Antecessor Price	0.032 (n.s)
		Profit Optimal Price	0.138*
		Base Case Price	-0.006 (n.s.)
	Volume Measurement	Relative Volume (b)	
		Absolute Volume	0.053 (n.s.)
	Item Level	SKU (b)	
		Brand	0.171***
	Method Used	Indirect Method (b)	
		Direct Method	0.206***

$r^2 = 0.205$, adjusted $r^2 = 0.166$, F = 5.29; degrees of freedom = 17, 350; p = 0.000

n = 368
standardized beta coefficients displayed
* p< 0.10, ** p< 0.05, *** p< 0.01
Dummies: type of brand***, price anchor n.s.

The explained variance is 20.5%. The model fit is satisfactory and highly significant (F = 5.29; degrees of freedom = 17, 350; p = 0.000, n = 368).

The most influential determinant is the quality of the product. For a price decrease, the price elasticities become stronger with increasing quality levels, indicating that consumers react more strongly when prices of products with higher quality levels are lowered. As this research does not assess temporary price changes such as price promotions, permanent demand is stimulated. Asymmetric switching to higher quality products when they reduce their prices has been well documented (Allenby/Rossi 1991; Blattberg/Wisniewski 1989; Kamakura/Russell 1989). Price reductions in higher quality products attract more consumers than price reductions in lower quality products (Allenby/Rossi 1991, p. 185).

The degree of differentiation is another influential factor, with higher levels of differentiation, consumers react less price sensitively to price reductions. Lower price elasticities are also found for premium and luxury brands compared to manufacturer brands.

The higher the level of complexity, the lower is the price elasticity. As products become more complex, the uncertainty increases due to the difficulty in understanding the product. This leads to higher procedural switching costs and therefore customers do not react that strongly to price reductions (Burnham/Frels/Mahajan 2003). Business-to-business transactions are characterized by lower price elasticities than business-to-consumer transactions. The higher the level of competition, the lower the price elasticity.

No significant influence can be found for the remaining product and market characteristics; i.e. the absolute price level, the stage of product life cycle, the price transparency and the frequency of purchase.

Looking at the research methodology, it becomes apparent that using a direct interview method leads to lower price elasticities compared to using indirect methods. The type of volume measurement, absolute vs. relative sales assessment, has no significant influence on the price elasticity. Price elasticities assessed at the brand level are lower than price elasticities assessed at the SKU level. When the profit-optimal price is used as the price anchor, price elasticities tend to be lower than when the current price serves as the anchor.

5.2.4.3 Analysis of Determinants - Price Elasticity for a Price Increase

In the next step, the determinants are analyzed using the price elasticities for a price increase as the independent variable in the regression analysis.

Chapter 5: Determinants of Price Elasticity 127

Looking at the price increase (figure 5-5), the explained variance is 23.8%, which is slightly higher than in the previous scenario. The model fit is satisfactory and highly significant (F = 6.43; degrees of freedom = 17, 350; p = 0.000, n = 368).

Figure 5-5: Results of Regression Analysis –
Determinants Analyzed with Price Elasticities for a Price Increase of 10%

	Determinants	Levels	Standardized β Coefficients
Product and Market Characteristics	Degree of Differentiation	7-point scale	0.318***
	Quality	7-point scale	-0.373***
	Type of Brand	Manufacturer Brand (b)	
		Premium Brand	0.123*
		Luxury Brand	0.236***
	Absolute Price Level	7-point scale	-0.199*
	Stage of Product Life Cycle	Maturity/Decline (b)	
		Introduction/Growth	-0.016 (n.s.)
	Level of Competition	7-point scale	0.025 (n.s.)
	Price Transparency	7-point scale	0.011 (n.s.)
	Level of Complexity	7-point scale	0.213***
	Frequency of Purchase	7-point scale	0.002 (n.s.)
	Type of Market	Business-to-Consumer (b)	
		Business-to-Business	0.070 (n.s.)
Research Methodology	Price Anchor	Current Price (b)	
		Antecessor Price	-0.055 (n.s.)
		Profit Optimal Price	0.123*
		Base Case Price	-0.090 (n.s.)
	Volume Measurement	Relative Volume (b)	
		Absolute Volume	0.124**
	Item Level	SKU (b)	
		Brand	0.155**
	Method Used	Indirect Method (b)	
		Direct Method	0.127*

r^2 = 0.238, adjusted r^2 = 0.201, F = 6.43; degrees of freedom = 17, 350; p = 0.000

n = 368
standardized beta coefficients displayed
* p< 0.10, ** p< 0.05, *** p< 0.01
Dummies: type of brand***, price anchor**

Overall, the influencing factors for a price increase are similar to the previous scenario of a price reduction. Higher quality also increases the level of price elasticity for a price increase. Buyers of low-priced brands could perceive a quality improvement in a price increase and therefore respond less strongly (Bemmaor 1984, p. 303). At a high level of quality buyers are more aware of product benefits and may use price less as a quality indicator (Bemmaor 1984; Klein/Leffler 1981), thus responding more strongly to a price increase.

The degree of differentiation lowers the price elasticity as does the level of complexity. It also holds true that premium and luxury brands have lower price elasticities than manufacturer brands. The level of complexity also significantly lowers the price elasticity. The higher the absolute price level, the higher the price elasticity. The price elasticities in the business-to-business market do not show a significant difference compared to the business-to-consumer market for a price increase; while for the average price elasticity and the price decrease the price elasticities in the business-to-business market are significantly lower. The price situation in the business-to-business sector is characterized by increased price pressure and declining prices (Diller 2004, pp. 947-968). Thus for price increases the customers react as price-sensitively as in the business-to-consumer market. The remaining product and market characteristics show no significant influence on the level of price elasticity.

Examining the research methodology, it becomes apparent that several aspects influence the price elasticity. When the research object is assessed at the brand level, the price elasticities are significantly lower compared to SKU level. In addition, the price elasticities are significantly lower when the profit-optimal price is used as a price anchor than when the current price is used. Again, it becomes apparent that utilizing direct interview methods lowers the price elasticities. Measuring the volume in absolute terms leads to lower price elasticities than measuring the volume in relative terms. Again, price elasticities measured at the brand level consistently display lower price elasticities.

5.2.4.4 Synthesis of Results

In figure 5-6 the results of the previous analyses are synthesized. The determinants were tested with the average price elasticity, the price elasticity for a price decrease and the price elasticity for a price increase. The effect on the magnitude of the price elasticity is shown, so the results and plus and minus signs are comparable to previous tables on the effect of determinants, for example in chapter 5.1.

Chapter 5: Determinants of Price Elasticity 129

Figure 5-6: Overview of Effects of Determinants on Magnitude of Price Elasticity

		Effect on Magnitude of Price Elasticity
Product and Market Characteristics	Degree of Differentiation	-
	Quality	+
	Type of Brand (vs. Manufacturer Brand)	Premium: -, Luxury: -
	Absolute Price Level	mixed results: + / n.s. / +
	Introduction/Growth (vs. Maturity/ Decline)	**n.s.**
	Level of Competition	mixed results: n.s. / - / n.s.
	Price Transparency	**n.s.**
	Level of Complexity	-
	Frequency of Purchase	**n.s.**
	B2B (vs. B2C)	mixed results: - / - / n.s.
Research Methodology	Price Anchor (vs. Current Price)	Ante: **n.s.**, Profit: -, Base: **n.s.**
	Absolute Volume (vs. Relative)	mixed results: - / n.s. / -
	Brand (vs. SKU)	-
	Direct Method (vs. Indirect)	-

- : negative impact on magnitude of PE
+ : positive impact on magnitude of PE
n.s.: not significant
bold font indicates consistent results across all three scenarios
otherwise the results are shown for:
average PE/PE for price decrease/PE for price increase

More than 80% of the determinants show consistent results in all three regression calculations. 59% to 65% of the tested determinants show significant effects (59% for PE price increase, 59% for PE price decrease, 65% for average PE). This is favorable compared to 40% of determinants in the meta-analysis of Bijmolt/Van Heerde/Pieters (2005) and 30% of the determinants in the meta-analysis of Tellis (1988) showing a significant effect on price elasticity.

Looking at the product and market characteristics, the following determinants consistently over the three scenarios lowered the price elasticity significantly: the degree of differentiation, the type of brand being premium or luxury brand and the level of complexity. The quality of the product consistently increased the price elasticity significantly.

No significant results were found consistently in all three scenarios for the stage of product life cycle, the price transparency and the frequency of purchase. The results were mixed for the absolute price level (increasing PE in two scenarios and n.s. in third), the level of competition (n.s. in two scenarios and decreasing PE in the third scenario) and the type of market (business-to-business market has lower PE in two scenarios and no significant effect in third scenario).

Regarding the research methodology, using the profit-optimal price as the price anchor leads consistently to significantly lower price elasticities, whereas using the antecessor price or the base case price leads consistently to no significant effect compared to using the current price as the price anchor. Also consistently across all three scenarios are the item level (brand vs. SKU) and the method used to derived the price response function (direct vs. indirect method). Using the brand vs. the stock-keeping unit in the research leads to lower price elasticities as does using direct methods to derive the price response function. The results are mixed for the volume measurement, using absolute volume measurement leads to lower price elasticities in two scenarios and to no significant effect in the third scenario.

5.2.5 Industry Specific Aspects

The influence of determinants might vary from industry to industry; therefore the determinants are also examined for various industry settings. In the industry analysis, the average price elasticity is examined. Overall, there are tremendous differences regarding determinants in an industry specific context.

Methodologically, stepwise-regression is used to identify the most relevant determinants in this more explorative setting. In stepwise regression, the independent variables are entered according to their statistical contribution in explaining the variance in the dependent variable. Researchers often apply stepwise regressions for sorting out the relative importance of determinants and for finding the most parsimonious set of variables that are most effective in predicting the dependent variable (Backhaus et al. 2006, pp. 105-111; Cohen 1991; Whittingham et al. 2006). Stepwise regressions are also used to reduce concerns about multicollinearity and to achieve acceptable levels of degrees of freedom given a limited sample size (Henning-Thurau/Houston/Heitjans 2009; Slater/Narver 2000). The procedure is also used in

Chapter 5: Determinants of Price Elasticity

marketing research to test results and is applied to identify the best subset of variables that predict a dependent variable (Schlosser/White/Lloyd 2006; Woodside 2012). In addition, the methodology is applied to test influencing factors on a dependent variable, e.g. business performance, in various industries (Yau et al. 2000). The procedure leads to the most influential variables for each industry. Another application is the identification of determinants, factors influencing a certain research subject at hand, e.g. marketing orientation or willingness to pay a premium (Ha-Brookshire/Norum 2011; Naudé/Desai/Murphy 2003).

As one of the goals in the research at hand is to identify the most influencing factors in the various industries in an exploratory way, this approach is suitable. The procedure leads to an economical model that is a rather complete but efficient description. It can also be understood as a procedure that searches for the set of variables that best explain the variation in the data.

The analysis of the determinants includes the same product categories that were analyzed regarding the magnitude of price elasticity. Also in line with the analysis of the magnitude of price elasticity, the same cut-off level on sample size is chosen and industries with more than 29 price elasticity cases are included in the analysis. Limiting the analysis to a specific industry automatically decreases the heterogeneity of the data set. In some industries several variables are constants and therefore cannot be used as predictors of the price elasticity within the industry setting. These cases will be pointed out in the analysis.

Due to the limited number of price elasticity cases, some variables also have to be taken out due to a very limited number of observations in order not to bias the overall analysis (Blankmeyer 2006). Because of the less heterogeneous data with several determinants being constant in the industry setting and also increased multicollinearity in various settings, the focus is on the more stable results and the input from the expert interviews is considered to provide insights and tendencies to be explored in further industry specific research with larger sample sizes. The following findings are therefore to be considered as first insights into a field currently not yet researched.

The industries are listed with decreasing sample size from 73 price elasticity cases for the automotive industry to 29 cases for pharmaceuticals and medical technology.

Automotive Industry

For the automotive industry (figure 5-7) the explained variance is 29.7% ($F = 9.70$; degrees of freedom = 3, 69; $p = 0.000$; $n = 73$).

Figure 5-7: Stepwise Regression Analysis – Automotive

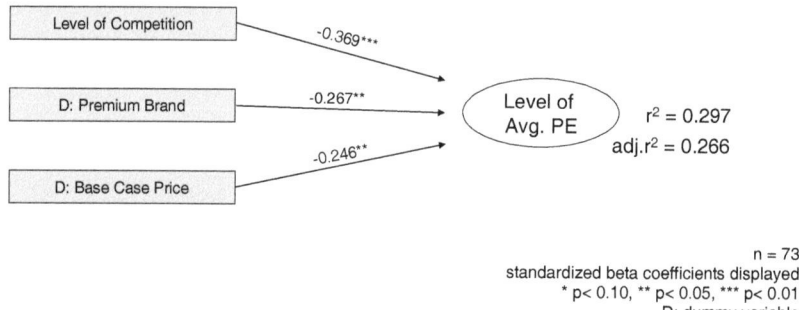

n = 73
standardized beta coefficients displayed
* p< 0.10, ** p< 0.05, *** p< 0.01
D: dummy variable

In the automotive industry the most influential factor is the level of competition, higher levels of competition lead to higher price elasticities. Expert interviews revealed that, for example, the introduction of inexpensive Chinese cars changes the elasticities in the low-end market. Before the decision was about buying or not buying a car, now a low cost alternative has been introduced.

Price elasticities are also higher for premium brands than manufacturer brands. If the data is limited to cars only, the premium brands have also higher price elasticities. However, in the data set at hand, the vast majority of brands are premium brands.

In the data set, the profit-optimal price was never used as a price anchor, therefore this variable was eliminated from the analysis. When the base case price is used as the price anchor, price elasticities tend to be higher; this has to be seen in comparison to the case when the current price is used as a price anchor. The base case price is the planned price for the product as determined by the manufacturer in the base case scenario.

Fast Moving Consumer Goods

The product category that is predominantly used in previous research, fast moving consumer goods, is examined. In contrast to the regression model for the overall data set, the explained variance of $r^2 = 0.749$ is substantially higher (F = 32.22; degrees of freedom = 5, 54; p = 0.000, n = 60). Figure 5-8 gives an overview of the influencing factors using the step-wise regression procedure.

Chapter 5: Determinants of Price Elasticity

Figure 5-8: Stepwise Regression Analysis – Fast Moving Consumer Goods

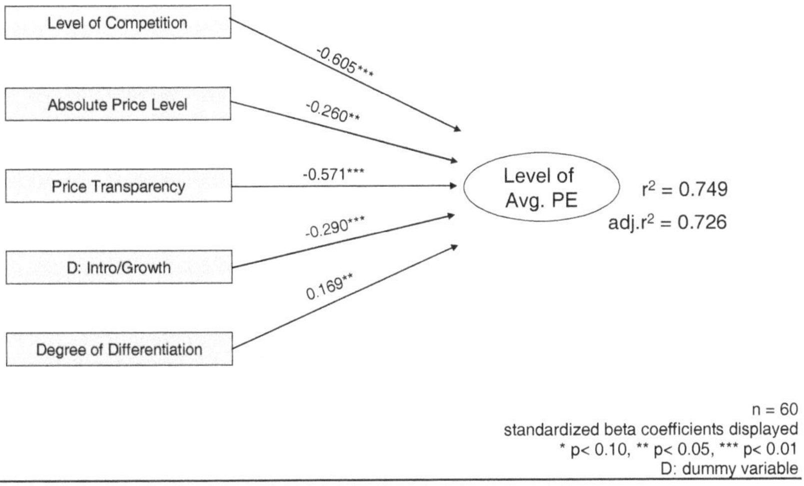

```
Level of Competition ──-0.605***──┐
                                   │
Absolute Price Level ──-0.260**──┐ │
                                  ↓↓
Price Transparency ──-0.571***──→ Level of    r² = 0.749
                                  Avg. PE     adj.r² = 0.726
D: Intro/Growth ──-0.290***──────↗↑
                                   │
Degree of Differentiation ──0.169**┘
```

n = 60
standardized beta coefficients displayed
* p< 0.10, ** p< 0.05, *** p< 0.01
D: dummy variable

Several variables are constant in this industry analysis or levels of dummy variables are non-existent and therefore have to be taken out of the analysis, e.g. the market is a business-to-consumer market, the research object was always modeled on the SKU level and the current price was always used as the price anchor.

The level of competition has a strong influence on the price elasticity. A higher level of competition leads to larger price elasticities. Previous research also confirms this relationship (Danaher/Brodie 2000; Narasimhan/Neslin/Sen 1996; Raju 1992). Even though research results are not always consistent, Bell/Chiang/Padmanabhan (1999) find a stronger reaction to price promotions in markets with a lower level of competition since this is somewhat novel in this environment compared to a highly competitive environment with frequent price promotions. Nevertheless, it is to be considered that the price changes examined in the research project at hand are, in contrast to most other price changes considered in previous projects, non-promotional, long-term price changes. The duration of the effect is, according to the most recent meta-analysis, short-term in 95 % of the time (Bijmolt/Van Heerde/Pieters 2005).

The price level has a positive effect on the magnitude of price elasticity. Customers are more price sensitive when purchasing more expensive items. More than 88% of the FMCG purchases in this dataset are however under the limit of 5 Euro.

Price transparency plays a role in the FMCG market, with a higher level of price transparency, the price elasticity increases. Overall, the variation in price transparency is limited in this category.

In the introduction and growth phase of the product life cycle, the elasticities tend to be higher than in the maturity and decline phase. Less experience with the product might lead to higher price elasticities in the early stages of the product life cycle and more experience with the product and stronger preferences in the later stages might make consumers more loyal and react less to price changes. This is in line with Bijmolt/Van Heerde/Pieters (2005, p. 147) who also found higher price elasticities for groceries in the introduction and growth phase.

The degree of differentiation lowers the absolute price elasticity as it does in the analysis for all price elasticity cases across all industries. Overall, the highest explanation of variance can be found in the traditionally used data set of fast moving consumer goods.

Industrial Goods

For industrial goods (figure 5-9), the situation is very specific most of the time; and generalizations are hard to be made regarding the managerial determinants. In this environment the interview methodology can explain 17.2 % of the variance (F = 11.24; degrees of freedom = 1, 54; p = 0.001, n = 56). Using direct interview methods in the primary research leads to lower price elasticities. Additional variables could not increase the explained variance. In this industry the dummy variable regarding the market characteristic business-to-business vs. business-to-consumer was excluded since all industrial goods are located in the business-to-business market. In addition, no luxury goods are part of the data set.

Figure 5-9: Stepwise Regression Analysis – Industrial Goods

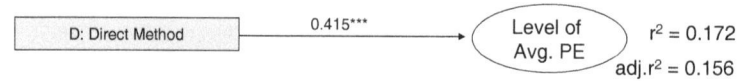

n = 56
standardized beta coefficients displayed
* p< 0.10, ** p< 0.05, *** p< 0.01
D: dummy variable

Chapter 5: Determinants of Price Elasticity 135

Away from Home Food

The away from home food market (figure 5-10) is a niche market and several variables are constants or have no correlations and are therefore automatically deleted from the regression analysis. The eliminated variables are the type of market, the level of complexity, the level of competition, the price transparency, the quality, the item definition SKU vs. brand, the method used to derive the price elasticity direct vs. indirect method, the volume measurement, the price anchor and the type of brand.

The determinants selected by the step-wise regression process, the absolute price level and the frequency of purchase, explain 36.4 % of the variance (F = 11.73; degrees of freedom = 2, 41; p = 0.000, n = 44). Consumers are more price sensitive at higher price levels. Products that are bought more frequently, display a lower price elasticity. Small sizes of soft drinks and coffee are bought more often at gas stations on freeways than other beverages and have relatively low price elasticities.

Figure 5-10: Stepwise Regression Analysis – Away from Home Food

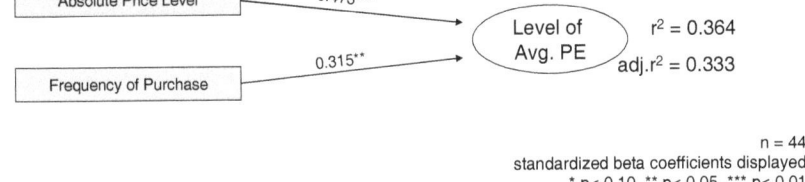

n = 44
standardized beta coefficients displayed
* p< 0.10, ** p< 0.05, *** p< 0.01

Logistics

Overall, the various industry analyses illustrate that the determinants differ considerably in the different industries. To provide some further illustrations, for logistics (figure 5-11), the frequency of purchase is an influential factor: with an increase in purchase frequency the price sensitivity of customers decreases. The explained variance is 16.5% (F = 5.720; degrees of freedom = 1, 29; p = 0.023, n = 31). Several variables are constant in this industry analysis or levels of dummy variables are non existent and therefore have to be taken out of the analysis. These are the item definition SKU vs. brand, the price anchors price of antecessor product and profit-optimal price as well as the variable luxury brand.

Figure 5-11: Stepwise Regression Analysis - Logistics

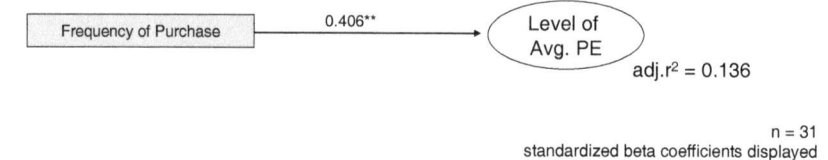

n = 31
standardized beta coefficients displayed
* p< 0.10, ** p< 0.05, *** p< 0.01

Consumer Durables

For consumer durables (figure 5-12) the price and the level of competition alone explain two thirds of the variance ($r^2 = 0.661$, F = 28.30; degrees of freedom = 2, 29; p = 0.000; n = 32). Again several variables have to be eliminated since they are constants, i.e. type of market, frequency of purchase, volume measure, stage of product life cycle, price anchor and type of brand.

As in the case for the other consumer goods, the price level influences the price elasticity positively, as does the level of competition. This is consistent with the notion that consumers are more price sensitive buying big ticket items than smaller ticket items (Bijmolt/Van Heerde/Pieters 2005).

However, in order to interpret the data, it has to be taken into account that the absolute price level is perfectly correlated with the positioning based on quality. Therefore, the price level could be replaced by quality and again, a higher quality would lead to higher price elasticities. In addition, the two dummy variables item definition on the brand level (vs. SKU) and using the direct method show a perfect negative correlation with the absolute price level. The multicollinearity of these variables is not a systematic problem but only arises in this specific scenario where the data is derived from three project sources and one project examines a product with a lower quality and price level that is modeled on the brand level using a direct interview method, while both other projects examine products with a higher quality and higher price level on the SKU level using an indirect interview method.

Chapter 5: Determinants of Price Elasticity 137

Figure 5-12: Stepwise Regression Analysis – Consumer Durables

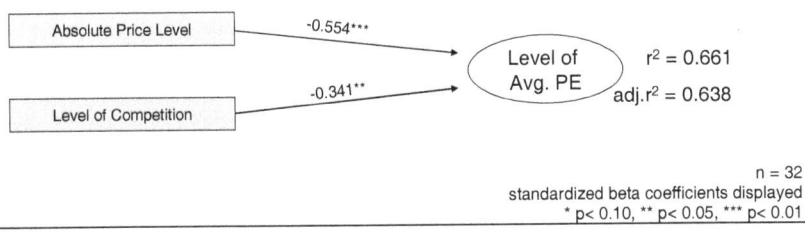

Pharmaceuticals and Medical Technology

The determinants were also explored for pharmaceuticals and medical technology. The explained variance is about two thirds (figure 5-13) (r^2 = 0.651; F = 24.25; degrees of freedom = 2, 26; p = 0.000; n = 29).

In this data set several variables had to be eliminated since they are constants: frequency of purchase, SKU vs. brand, volume measurement, stage of product life cycle, and the luxury brand. The price anchor dummy variable had to be adjusted. The data contained either the profit-optimal price as the price anchor or the price of an antecessor product as the price anchor.

Figure 5-13: Stepwise Regression Analysis – Pharmaceuticals and Medical Technology

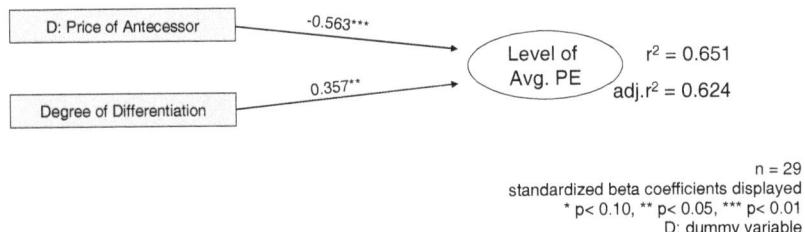

The price anchor used in the primary research plays an important role. If the price of an antecessor product is used, which is most likely the case for next generation products, the price elasticity tends to be higher. The reference of an antecessor product has a strong influence on the price elasticity and decision makers tend to become more price sensitive. The dummy variable has to be interpreted against the profit-optimal price since those are the only two price anchors used in this data subset.

If the product is nonetheless able to differentiate itself from other treatment options, the price elasticity decreases. This is in line with various project examples provided by the industry experts. The first substantially improved versions of a medical product x with new features could charge a high price premium. There was almost no price elasticity for this product. In addition, an increase in marketing efforts such as journal advertising, organization of meetings and events causes price elasticities to decrease in the pharmaceutical industry (Narayanan/Desiraju/Chintagunta 2004). However, marketing expenditures do not necessarily reduce the price elasticity as demonstrated in a study in the Netherlands (Leeflang/Wieringa 2010). This demonstrates that country-specifics like health-insurance coverage and regulations of pharmaceutical prices have to be taken into account when assessing price elasticities. With almost perfect health coverage such as in the Netherlands there is no strong financial stimulus for patients or physicians to be price sensitive (Leeflang/Wieringa 2010, p. 131).

5.3 Summary

In order to get a better understanding of determinants of price elasticity, the academic data was reviewed to see what determinants were actually studied and what additional insights can be gained from the data. The review discovered that most studies do not analyze determinants of price elasticity as the focus of their research, often it is only a by-product. Only very few studies specifically list coefficients and significance levels of determinants. Some studies examine competitive effects through cross-price elasticities and the own-price elasticities are only reported in a side note. The focus is on research methodology comparing different models and data sources. Therefore, most determinants of the meta-analysis of Bijmolt/Van Heerde/Pieters (2005) were never directly analyzed in the original study. Additional analyses were performed, e.g. correlations of market shares and price elasticities to gain further insights.

The following determinants could be analyzed based on the academic publications: market share, level of competition, premium positioning/quality, brand ownership, direction of price change and customer characteristics. In the second data source, consulting project data, additional determinants could be analyzed. The key results for all determinants are presented in the following summary.

MARKET, PRODUCT AND CUSTOMER CHARACTERISTICS

The following determinants belong to either market, product or customer characteristics.

Market Share

The analysis of the market share shows that large share brands tend to have lower price elasticities than brands with a smaller market share. This can be explained by the mathematical calculation of price elasticity. The consulting project data has limited information on market share of brands but a descriptive analysis of this data supports this tendency.

Level of Competition

The level of competition is studied via several measures within the academic data and seems to explain some variance. The results are not congruent but lower levels of competition can lead to reduced price elasticities. In the consulting project data the level of competition has no significant effect for the overall scenario, but with a higher level of competition the price elasticities are stronger in the fast moving consumer goods, consumer durables and automotive categories.

Premium Positioning and Quality Type of Brand

In the academic studies, premium positioning and quality aspects are not distinguished. The consulting project data differentiates between the aspect of the branding on the one hand and the quality level on the other hand. The academic literature indicates higher price elasticities for higher quality products. For premium brands the results are mixed. Looking at the overall consulting project data, it becomes apparent that higher quality products tend to have higher price elasticities. Nevertheless, premium and luxury brands have lower price elasticities than manufacturing brands. In the automotive industry the premium brands have, however, higher price elasticities than the manufacturing brands.

Brand Ownership

The brand ownership private label vs. manufacturer brand led to mixed results in the academic data, sometimes the private label has the lowest price elasticity, sometimes the highest and sometimes it ranges among the other brands. Bijmolt/Van Heerde/Pieters (2005) also found no significant effect of the brand ownership. The consulting project data did not contain any private labels, therefore this aspect could not be analyzed.

Direction of Price Change

The direction of price change was not explicitly studied in the academic data set, but Moon/Russell/Duvvuri (2006) found higher price elasticities for a price increase vs. a price decrease. In the consulting project data, a comparison of the magnitudes of price elasticities shows a significant difference between the means of the price decrease and increase. The mean price elasticity for the price decrease is significantly lower than the price elasticity for the price increase ($p < 0.01$). The weaker reaction to price decreases is also indicated by the lower median of -1.07 vs. -1.50 for the increase. The direction of the price change was not modeled as a determinant in the regression analyses since the focus of this research was on the average price elasticities; and then the determinants were assessed for a price decrease and a price increase separately.

Customer Characteristics

Customer characteristics are extensively studied in the academic publications and an impact on the price elasticities is demonstrated in several studies, the results are however mixed even when the same determinants are studied in the same market. Generalizations are limited if possible at all. Different studies lead to opposite results on the effect of determinants. Customer characteristics were not studied in the consulting project data due to limited data availability and the chosen focus on product and market characteristics.

Degree of Differentiation

The degree of differentiation is one of the most important influencing factors on the price elasticity. It lowers the price elasticities in the overall data set and is a very influential factor in the pharmaceutical and medical technology market.

Absolute Price Level

The absolute price level of a product increases the level of price elasticity for the average price elasticity. Looking at a price increase this effect also increases the price elasticity, but the effect is not significant for the price decrease. In the industry analysis, however, a higher price level increases the price elasticities in the fast moving consumer market, the away from home food market and the consumer durable market.

Stage of Product Life Cycle

The stage of product life cycle shows no significant effect on the price elasticities in the overall data set but higher price elasticities in the early stage of the product life cycle can be found for fast moving consumer goods.

Price Transparency

The price transparency only shows a significant effect in the fast moving consumer goods market. The price elasticity increases with higher levels of price transparency. The variation in price transparency is, however, rather limited in this industry.

Level of Complexity

A higher level of complexity leads to lower price elasticities in the overall data set. Within a certain industry, the level of complexity was rather similar therefore no further insights could be gained in the industry analysis.

Frequency of Purchase

The frequency of purchase is significant in the away from home fast food and snack market and in the logistics industry. Products that are more frequently bought display lower price elasticities.

Type of Market

The type of market (business-to-business vs. business-to-consumer market) has a significant influence on the price elasticities, with lower price elasticities for the business-to-business market, in the overall data set for the average price elasticity and the price decrease but not for a price increase. The effects are, however, weak. In the industry analysis this variable is usually a constant and therefore excluded from the analysis.

RESEARCH METHODOLOGY

The remaining determinants belong to the research methodology.

Price Anchor

The price anchor used to calculate the price elasticities has an influence on the price elasticities in various cases. Using the profit-optimal price as the price anchor leads to lower price elasticities in the overall data set. The base case and the price of the antecessor product consistently show no significant effects.

Using the price of an antecessor product in the pharmaceutical industry increases the price elasticity significantly vs. using the profit-optimal price. Only these two price anchors are used in the pharmaceutical subset. In the automotive industry, using the base case price significantly increases the price elasticity.

Volume Measurement

Using absolute volume measurement, more specifically asking respondents about the sales/buying volume vs. asking them about relative assessments of volume (market shares or indices) shows a small and weak effect in the average price elasticity scenario whereas using absolute volume leads to lower price elasticities. The effect can also be found for a price increase but is not significant for a price decrease.

Item Level

Whether the price elasticity is assessed at the SKU or brand level has a significant influence in the overall data set. The price elasticities for brands are significantly lower than those for stock-keeping units, which is in line with the results of Bijmolt/Van Heerde/Pieters (2005).

Method Used to Derive Price Response Function

Using a direct interview method leads to lower price elasticities compared to using indirect methods. This is consistently true for all analyses in the overall data set. In the industry specific analyses this effect can be found for industrial goods.

CONCLUSIONS

Overall, the most important determinants in the consulting project data are the degree of differentiation and the positioning based on quality. If only these two variables are included in the regression model (figure 5-14), the explained variance is 14.8% (F = 32.78, degrees of freedom = 2, 378, p = 0.000, n = 381). The degree of differentiation has a negative effect on the magnitude of price elasticity. The more a product is able to differentiate itself, the lower are the price elasticities. Consumers do not have alternatives in their mindset and are therefore bound to the product. The quality level increases the price elasticity. Higher quality products tend to have higher price elasticities.

Figure 5-14: Regression Analysis – Key Determinants

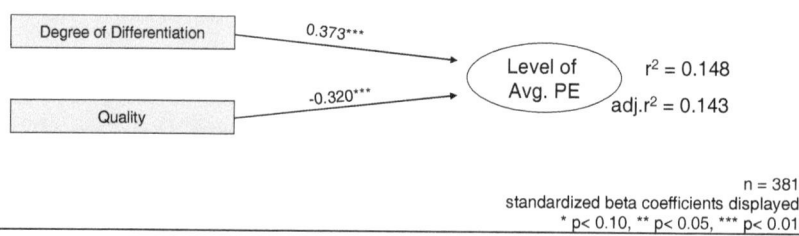

n = 381
standardized beta coefficients displayed
* p< 0.10, ** p< 0.05, *** p< 0.01

The most influential determinants vary in the industry settings; the variation in the industry specific data set is more limited and several determinants are usually constants in an industry specific setting and thus limit the analysis options.

In addition, it is extremely difficult to generalize determinants, especially given the variation in magnitude of price elasticity described in chapter 4. This also reflected in previous meta-analyses (Tellis 1988; Bijmolt/Van Heerde/Pieters 2005) and other previous research. Taking the studies of Montgomery (1997) and Hoch et al. (1995) as an example, since both studies examine the same customer characteristics and competitive characteristics in the fast moving consumer goods market, they find opposite directions of effects for some determinants and can explain 22% of the variance for orange juice, 34% for toothpaste and 86% for frozen dinners, which illustrates that even the same variables in the same setting have a very different explanatory power.

6 Conclusions

In this chapter the key findings of the research project are addressed. Then the implications for research, including limitations of the research und suggestions for future research directions are discussed. In the last step implications for management are addressed.

6.1 Key Findings

Given the high importance of the price management within the marketing field and the special role of price within the marketing mix, a highly relevant topic both to academic research and managerial practice is addressed in this research (Leone et al. 2012; Monroe 2003; Simon/Fassnacht 2009). Price elasticities are a useful measure to compare price sensitivities across products and markets (Sivakumar 2001, p. 1). Understanding this key figure enhances managerial decision making to set profit-optimal prices. Managers need to be aware that the price has the strongest impact of all marketing variables on profitability (Sethuraman/Tellis/Briesch 2011; Simon/Fassnacht 2009, p. 3). Making the right pricing decisions is the fastest and most effective way to grow profits (Baker/Marn/Zawada 2010, p. 3).

Current research on price elasticities is primarily based on scanner data and fast moving consumer goods (Bijmolt/Van Heerde/Pieters 2005). An assessment of a broader product spectrum is currently lacking. It is not only important to understand the magnitude of the price elasticity but it is also important to understand the determinants of price elasticity. Knowledge about both the magnitude and the determinants allows understanding markets and influencing prices and determinants in a profit-optimal way.

In studying additional products and determinants, the research at hand could expand the current knowledge. A high degree of managerial relevance was the motivation of this research as there is a current lack in perceived relevance and understanding of price elasticities in the managerial vs. the academic field (Simon/Fassnacht 2009, p. 10).

Against this background and the academic and managerial need to enhance the knowledge on price elasticities, this research had two main objectives:

- The **first objective** was to expand the knowledge on the magnitude of price elasticity.

- The **second objective** was to expand the knowledge on determinants of price elasticity.

Chapter 6: Conclusions

The research objectives were addressed in the following, subsequent steps:

In the first step, the foundation of the research was built (chapter 2). The conceptual and theoretical background was established via definitions and explanations of the price elasticity concept and its relation to the price response functions and price optimization. In addition, the terminology in the price elasticity context and related concepts were discussed. Furthermore, prior research on the magnitude of price elasticity and the determinants of price elasticity was presented.

In the next step, two data sets were created, one derived from academic publications and one from consulting project data (chapter 3). The data collection followed specific criteria to ensure a high level of data quality and to facilitate comparability. The author thoroughly examined each academic study and each consulting project and collected the research focus, methodology and other relevant information on price elasticities that was available. For the consulting project data, the price elasticities were calculated with a consistent methodology. Additionally, overviews of the selected studies and projects were provided.

Having built these data sets, the research could then address the main objectives to enhance the knowledge on the magnitude (chapter 4) and the determinants of price elasticity (chapter 5).

Key Findings on Magnitude of Price Elasticity

The academic data set was analyzed overall and in contrast to previous research also for specific product categories (chapter 4). This enabled the author to show frequency distributions for a variety of products and compare descriptive data. The graphical illustrations that were provided facilitate the interpretation of the price elasticity data and demonstrate various patterns and differences in the frequency distribution.

Table 6-1 gives an overview of the mean price elasticities for the academic data set. The overall mean is -2.51, the mean price elasticities for the analyzed product categories vary between -4.09 for bathroom tissues and -1.21 for shampoo. The product categories are listed in order of decreasing magnitude of price elasticity.

Table 6-1: Mean Price Elasticities – Academic Data

Product Category	# of Price Elasticities	Mean	Product Category	# of Price Elasticities	Mean
All	863	-2.51	Yogurt	64	-2.75
Bathroom Tissues	66	-4.09	Automotive Parts	66	-2.22
Tuna	33	-4.00	Laundry Detergent	127	-1.93
Coffee	46	-3.54	Saltine Crackers	30	-1.44
Margarine	62	-3.26	Peanut Butter	61	-1.37
Ketchup	80	-3.03	Shampoo	39	-1.21

The wide spread in magnitude within each product category was illustrated in chapter 4. Furthermore, not only products but also brands and brand sizes were compared across studies; this enabled the author to go a step deeper into the analysis. This additional analysis showed that even for the same brand there is a wide variety in measured price elasticities in the academic data.

Beyond the academic data, a unique set of price elasticities based on consulting project data was calculated. This additional data source stemming from survey data was also analyzed regarding the magnitude of price elasticity. A broad spectrum of products across various industries was covered in this data.

Table 6-2 gives an overview of the mean price elasticities found in the consulting project data. The overall mean is -1.73, the mean price elasticities for the analyzed product categories vary between -3.41 for food consumed away from home (primarily snacks and fast food consumed while travelling) and -1.14 for fast moving consumer goods. Again, the product categories are listed in order of decreasing magnitude of price elasticity.

Table 6-2: Mean Price Elasticities – Consulting Project Data

Product Category	# of Price Elasticities	Mean
All	386	-1.73
Away from Home Food	44	-3.41
Automotive	73	-2.02
Logistics	36	-1.62
Industrial Goods	56	-1.54
Consumer Durables	32	-1.40
Pharma and Medtech	29	-1.33
FMCG	60	-1.14

The consulting project data provided the opportunity to compare price elasticities for a price increase with price elasticities for a price decrease. The mean of the price decrease (-1.62) is significantly lower ($p < 0.01$) than that of the price increase (-1.84). The median illustrates the stronger reaction to price increases even better (-1.50 vs. -1.07) since it is less influenced by outliers.

The mean price elasticity of the academic data set is -2.51 (median = -2.21; SD = 1.81). This is very much in line with the mean of Bijmolt/Van Heerde/Pieters (2005) of -2.62 (median = -2.22, SD = 2.21). The mean of the consulting project data at -1.73 is substantially lower (median = -1.29, SD = 1.33) but close to Tellis' (1988) mean of -1.76 (SD = 1.74). Tellis' (1988) database contains more diverse data with a larger percentage of durables and pharmaceuticals than Bijmolt/Van Heerde/Pieters (2005) and the academic data set that contains no durables and pharmaceuticals for the last 25 years of research. This could also contribute to the fact that the descriptive values of the consulting project data are closer to Tellis (1988). In contrast to the scanner based data in the previous meta-analysis which primarily capture short-term price changes, the research at hand focuses on long-term price changes and regular prices. Short-term, promotional price elasticities are known to be substantially higher than long-term, regular price elasticities (Simon/Fassnacht 2009, p. 498).

Key Findings on Determinants of Price Elasticity

Six categories of determinants were identified that could be used to gain insights from in the academic publications. Table 6-3 shows the key results.

Table 6-3: Categories of Determinants Identified in Academic Publications

Determinant	Key Results of Academic Publications	Impact on Price Elasticity
Market Share	the majority of studies (8 out of 11 studies) find a negative impact on the magnitude of PE, i.e. the higher the market share, the lower the PE	-
Level of Competition	no consistent operationalization of the level of competition, results are mixed	+/-
Premium Positioning/Quality	the premium positioning and the quality aspect are not clearly differentiated in the academic studies premium brands: mixed results high quality brands tend to have higher PE	+/- +
Brand Ownership	there is no clear tendency of the PE of private labels 4 studies: +, 4 studies: -, 2 studies: no impact observed	+/-/o
Direction of Price Change	2 studies analyze opposite direction of price change on PE one study finds higher PE for price increases the second study reports not details on the data	+/?
Customer Characteristics	a wide variety of customer characteristics are tested in 8 studies, the results are rather mixed	+/-/o

- : negative impact on magnitude of PE
+ : positive impact on magnitude of PE
o: no impact observed
?: unknown impact due to lack of data

In the consulting project data two categories of determinants could be tested. Market and product characteristics belong to the first category and research methodology to the second category tested. The determinants were assessed overall and in an explorative way for various industry settings. Table 6-4 provides an overview of the results.

Chapter 6: Conclusions 149

Table 6-4: Determinants of Price Elasticity – Consulting Project Data Product and Market Characteristics

Determinant	Key Results of Consulting Project Data	Overall Impact on Price Elasticity
Degree of Differentiation	products with a higher degree of differentiation tend to have lower PE most influential in pharma and medical technology	-
Quality	higher quality products tend to have higher PE	+
Type of Brand	premium and luxury brands tend to have lower PE than manufacturer brands, no data on private labels automotive: premium brands tend to have higher PE than manufacturer brands	-/+
Absolute Price Level	overall data set: weak effect +/n.s./+ For consumer goods the higher the absolute price level, the higher the PE + FMCG +, away from home food +, consumer durables +	+
Stage of Product Life Cycle	overall data set: n.s. FMCG: early stages PE > late stages	n.s./+
Level of Competition	overall data set: n.s./-/n.s. FMCG + consumer durables + automotive +	n.s./-/+
Price Transparency	overall data set: n.s. FMCG +	n.s./+
Level of Complexity	overall data set: products with higher level of complexity tend to have lower PE	-
Frequency of Purchase	overall data set: n.s. away from home food: - logistics: -	n.s./-
Type of Market	overall data set: -/-/n.s. business-to-business markets tend to have lower PE than business-to-consumer markets	-

- : negative impact on magnitude of PE
+ : positive impact on magnitude of PE
n.s.: not significant

The second determinant group tested is the research methodology. Table 6-5 gives an overview of key research results.

Table 6-5: Determinants of Price Elasticity – Consulting Project Data Research Methodology

Determinant	Key Results of Consulting Project Data	Overall Impact on Price Elasticity
Price Anchor	overall data set : profit-optimal price PE < current price PE, antecessor price n.s., base case price n.s. automotive: base case price > current price pharma and medtech: antecessor product price > profit-optimal price	- n.s. +
Volume Measurement	overall data set: -/n.s./- absolute volume measurement tends to yield lower PE than relative volume measurement	-
SKU vs. Brand	brands tend to have lower PE than SKU	-
Method Used	using direct method tends to yield lower PE than indirect method	-

- : negative impact on magnitude of PE
+ : positive impact on magnitude of PE
n.s.: not significant

Overall, the most influential variables are the degree of differentiation and the positioning based on quality (cf. figure 5-7 in chapter 5). These two determinants explain 14.8% of the variance. The degree of differentiation lowers the price elasticity, while the level of quality increases the price elasticity.

6.2 Implications for Research

The implications of this research are divided in two aspects. The first aspect focuses on the contributions of this work for the academic field. The second aspect addresses limitations of this work and future research directions.

This work extends prior research in various ways:

Chapter 6: Conclusions

- First, a refined set of the academic data was assessed and the analysis went beyond the previous meta-analytic research by analyzing price elasticities more in detail for various product categories across studies. Price elasticities were not only compared on a product level but also compared on a brand and brand size level. Therefore, the research provides new meta-analytic insights on the magnitude of price elasticities.

- In addition to the academic data a second data source was analyzed – consulting project data. The focus was laid on price elasticities derived from survey data. This data provided the opportunity to extend the knowledge on price elasticities based on a previously underutilized data source (Miller et al. 2011, p. 182). Previous price elasticity analyses stem primarily from scanner data. As research is asking for a high degree of managerial relevance, this aim is ensured by analyzing data that was used in real business cases when actual companies asked for pricing advice regarding their products or services.

- The comparability of price elasticity data was enhanced by using a more consistent estimation method than previous research. In addition, the project data was sourced from a single consulting company that adheres to the same quality standards and methodology in primary research, which facilitates comparability.

- Determinants of price elasticities were explored beyond previously assessed variables. Academic studies were examined to find out what determinants were explicitly and indirectly studied. Beyond that, previous research was extended by exploring additional determinants in the consulting project data. Another contribution of the research at hand is that determinants were not only analyzed for the overall data set but also various industry settings were explored.

- A broader range of products and more diverse industries were covered in this research than in previous research which primarily had covered fast moving consumer goods. The current research project provides a detailed overview on price elasticities for a broad product spectrum covering not only business-to-consumer markets but also business-to-business markets. Therefore, the information on price elasticity currently available is extended.

- The focus on regular prices and long-term price changes addressed another gap in research, as previous price elasticity research is primarily based on short-term price changes and price promotions.

In the next passage limitations of this research are pointed out and future research directions are suggested:

- One limitation is that the coding procedure for the determinants in the consulting project data set was not performed by two independent coders as this was not viable in the data setting. Coders must have considerable background in the methodology and the specific research domain at issue to perform the coding task well, they must not only understand the coding protocol in detail and depth but must have the knowledge and skills to properly read and interpret the data reports (Lipsey/Wilson 2001, p. 88). Ideally, the coding procedure should be conducted by two independent coders and inconsistencies in coding should then be resolved in a discussion by the two coders or by an independent third judge. This procedure provides a test of reliability checking the agreements of coders. The coding of this research was performed by the author who made herself familiar with the product or service by reading background information on the project, final project reports and discussions with the project members. In cases where access to the data was restricted due to confidentiality agreements or where the author lacked the necessary industry expertise, the coding was discussed with industry experts. The expert provided their assessment based on a detailed coding sheet and discussions with the author. Generally, the coder agreement and coder reliability are typically quite good, even for more complex coding tasks (Cooper 1998, pp. 95-97). It is common for smaller studies that the coding is entirely done by the researcher (Lipsey/Wilson 2001, p. 90).

- Another limitation of the consulting project data is that is has a bias towards high quality brands. The data set contains no private labels or inferior products. Looking at the quality level, 98% of the data ranges on the quality level standard, high and premium quality (level 4 to 6 on the 7-point scale). The data contains no cases in the very low and the low quality level and just very few cases in the mass market quality und luxury quality level. A potential explanation for this fact could be in the sampling of firms hiring a consulting company and making the investment to optimize their pricing strategy. Not all companies are willing or in a position to make this investment. In future research, it is recommended to assess the findings in a broader data set including private labels and more luxury brands.

- The industry analyses consist of a smaller sample sizes (n = 29 to n = 73), therefore, further industry specific quantitative analyses on price elasticities are recommended with larger sample sizes. Some industries are too small for further analyses, in other industries price changes in only one direction were analyzed and therefore they were excluded in this analysis. In general, there is potential to go deeper in the industry specific exploration in future research.

- In this research, the consulting project data was assessed with multiple linear regression analysis, including all variables and also with a step-wise regression procedure. This methodology is used to get a first understanding of the newly available data source and its determinants. Building on this analysis, it is recommended to assess the data with further statistical methodologies. As seen in the academic data, the estimation of price elasticities varies strongly depending on the estimation model. Further analyses of the determinants will lead to valuable insights on the determinants of price elasticity.

- Some concerns about the hypothetical nature of using survey data can be raised, since it measures consumers' hypothetical, rather than actual, purchase behavior and thus can generate a bias. Nevertheless recent research demonstrates that this can still lead to the right price response functions (the basis for calculating the price elasticity) and therefore the right decisions about the optimal price level (Miller et al. 2011, p. 179). "Researchers focusing on the mean accuracy of hypothetical approaches may have underestimated their value in guiding managerial decision making" (Miller et al. 2011, p. 182).

6.3 Implications for Management

This research has a number of key learning points and implications for managerial practice. The motivation of the research was also founded in the current gap in knowledge between academia and management.

The most important point for managers is to understand that the price elasticity for a certain product is very specific and hard to generalize. Managers involved in price-setting have to understand their market and brands as well as the factors influencing the price elasticity. This research supports Gabor's view that "there is no such thing as *the* price elasticity of demand" (Gabor 1988, p. 18). For price managers, it is not enough to look at academic publications to estimate the price elasticity of a product in order to set a profit-optimal price. The current analysis shows the wide variation in price elasticities estimated for a specific product category and even for brands and brand sizes. A broad range of price elasticities is not specific enough to provide the necessary information to optimize the price setting. Therefore, a specific price elasticity assessment is recommended with a methodological approach that is well understood since the chosen methodology can influence the price elasticity.

Price elasticities are not constant but can change over time (Parker/Neelamegham 1997; Sethuraman/Tellis/Briesch 2011; Simon 1979), therefore regular monitoring is suggested. "Most business firms do not make sufficient efforts to track the sensitivity of demand for their products to price changes and to price differences over time" Monroe (2003, p. 155). The investment in market research is recommended to optimize the price. Each market situation is unique, so generalizations are very hard to make and an individual analysis is recommended to capture the specific market situation.

"Price elasticities are notoriously difficult to measure" (Ramirez/Goldsmith 2009, p. 199). Price elasticities are considered to be a very useful metric for 39% of senior marketing managers, the optimal price is very useful for 41% of senior marketing managers; both percentages are relatively low compared to the most important metric net profit which 91% of the respondents considered to be very useful (Bendle et al. 2010, p. 20). However, managers need to better understand the relationship between price and net profit. The price has a very strong and direct influence on net profit, much stronger than other marketing variables (Simon/Fassnacht 2009, p. 5; Sethuraman/Tellis/Briesch 2011). It should be also kept in mind that price elasticity measures the reaction to a price change, not the willingness to pay for a certain product and not the demand at a given price point.

Understanding the price elasticity will allow price settings according to the company's objectives. The market environment and influencing factors have to be understood since they vary tremendously. The determinants of price elasticity also vary significantly and need to be understood in the specific context. In general, it is suggested that managers aim to increase the degree of differentiation of the product or service and thus lower the price sensitivity of consumers. Especially for high quality products, setting the right price point is very important since price changes cause substantial differences in sales and therefore profit.

It is recommended that managers evaluate the pricing environment and are bold enough to raise prices if this will increase profitability. Many managers fear losing market share, however, it is advised to manage for profit not market share (Simon/Bilstein/Luby 2006). To avoid price increases the tactic of reducing content is used, e.g. reducing the pack size from 32oz to 28 oz. In the past content reductions worked since they were often not noticed but consumers become more aware of these practices and consumers with more knowledge on these pricing tactics perceive content reduction as less favorable versus price increases and this leads to less favorable attitudes towards the brand (Kachersky 2011). Also changes in shipping charges might cause much higher price elasticities than changes in product prices (Smith/Brynjolfsson 2001). In partioned prices, the components with lower consumption benefits have higher price elasticities (Hamilton/Srivastava 2008). This provides insights for management to reflect the value of the product in the pricing.

Chapter 6: Conclusions

In general, it is suggested that managers focus more on permanent prices than short-term price promotions. The long-term profitability of price promotions is questionable (Ailawadi/Neslin/Gedenk 2001). For certain products, consumers tend to wait for a deal to buy but do not consume more overall, e.g. laundry detergent (Ailawadi et al. 2007). Due to limited data Bijmolt/Van Heerde/Pieters (2005) did not extensively assess long-term effects (5% of cases) and regular prices (1% of cases). This research finds lower price elasticities for long-term regular price changes than previous academic research results that focused on short-term effect and promotional price changes.

Overall, the findings of this research indicate that price is a potent instrument in the marketing mix of many industries. This underscores the importance of continuing price elasticity research not only in the academic field but also in managerial practice.

Appendix

Appendix A:	Top 25 Marketing Journals	157
Appendix B:	Price Elasticity Database – Overview of All Cases	159
Appendix C:	Selected Academic Studies on Price Elasticities – Further Information	160
Appendix D-1:	Expert Interviews - Evaluation of Product Characteristics, Hypothesized Effect on Price Elasticity, Level of Agreement and Expected Strength of Effect	166
Appendix D-2:	Expert Interviews - Evaluation of Product Characteristics, Agreement with Hypothesized Effect on Price Elasticity	166
Appendix D-3:	Expert Interviews - Evaluation of Market Characteristics, Hypothesized Effect on Price, Level of Agreement and Expected Strength of Effect	167
Appendix D-4:	Expert Interviews - Evaluation of Market Characteristics, Agreement with Hypothesized Effect on Price Elasticity	167
Appendix D-5:	Expert Interviews - Evaluation of Customer Characteristics, Hypothesized Effect on Price Elasticity, Level of Agreement and Expected Strength of Effect	168
Appendix D-6:	Expert Interviews - Evaluation of Customer Characteristics, Agreement with Hypothesized Effect on Price Elasticity	168

Appendix A: Top 25 Marketing Journals

	Bauerly/Johnson (2005) Syllabi Analysis (109 syllabi)	Hult/Neese/Bashaw (1997) Doctoral Survey (118 doctoral faculty)	Hoch et al. (1997) Overall Survey (309 faculty)	Baumgartner/Pieters (2003) Citation Analysis (49 journal, 33,226 citations in 1996-1997)
1	Journal of Marketing	Journal of Marketing	Journal of Marketing	Journal of Marketing
2	Journal of Consumer Research	Journal of Marketing Research	Journal of Marketing Research	Journal of Marketing Research
3	Journal of Marketing Research	Journal of Consumer Research	Journal of Consumer Research	Journal of Consumer Research
4	Marketing Science	Marketing Science	Journal of Retailing	Harvard Business Review
5	Journal of the Academy of Marketing Science	Journal of Retailing	Journal of the Academy of Marketing Science	Management Science
6	Management Science	Journal of the Academy of Marketing Science	Marketing Science	Advances in Consumer Research
7	Journal of Personality & Social Psychology	Harvard Business Review	Harvard Business Review	Marketing Science
8	Psychological Bulletin	Management Science	Journal of Business Research	Journal of the Academy of Marketing Science
9	Strategic Management Journal	Journal of Business Research	Journal of Advertising	Journal of Retailing
10	Harvard Business Review	Journal of Advertising	Journal of Advertising Research	Industrial Marketing Management
11	Psychological Review	Journal of Advertising Research	Management Science	Journal of Advertising Research
12	American Psychologist	Journal of Public Policy & Marketing	Journal of Personal Selling and Sales Management	Journal of Business Research
13	Journal of International Business Studies	Sloan Management Review	Advances in Consumer Research	Journal of International Business Studies
14	Academy of Management Review	Journal of Business	Journal of Public Policy & Marketing	Sloan Management Review
15	International Journal of Research in Marketing	International Journal of Research in Marketing	Journal of Marketing Education	Journal of Advertising
16	Annual Review of Psychology	Psychology & Marketing	Psychology & Marketing	Journal of Product Innovation Management

Rank	Bauerly/Johnson (2005) Syllabi Analysis (109 syllabi)	Hult/Neese/Bashaw (1997) Doctoral Survey (118 doctoral faculty)	Hoch et al. (1997) Overall Survey (309 faculty)	Baumgartner/Pieters (2003) Citation Analysis (49 journal, 33,226 citations in 1996-1997)
17	Administrative Science Quarterly	Advances in Consumer Research	Sloan Management Review	European Journal of Marketing
18	Journal of Business	Journal of Personal Selling and Sales Management	Journal of Business	Journal of Personal Selling and Sales Management
19	Structural Equation Marketing	Industrial Marketing Management	Journal of International Business Studies	California Management Review
20	American Economic Review	Decision Sciences	Industrial Marketing Management	Business Horizons
21	Journal of Retailing	Journal of Consumer Psychology	Journal of Consumer Marketing	Journal of Public Policy & Marketing
22	Advances in Consumer Research	Journal of International Business Studies	California Management Review	International Journal of Research in Marketing
23	Marketing Letters	Marketing Letters	Business Horizons	Journal of Business Ethics
24	Journal of Services Marketing	California Management Review	Journal of International Marketing	Journal of Marketing Education
25	Academy of Management Journal and Journal of Consumer Psychology	Journal of Product Innovation Management and Journal of International Marketing	Journal of Services Marketing	Marketing Letters

Source: Adapted from Bauerly/Johnson 2005, p. 323

Appendix B: Price Elasticity Database – Overview of All Cases

Product category	Total	Product category	Total
food & beverages	58	automotive spare parts	6
automobile	42	copper crim fitting	6
telecommunication / - package	28	technology	5
medication	27	tunnel toll	5
automotive tires	26	bed mattress	5
parcel services	25	cranes and crane parts	4
dermal filler	20	insurance	4
self-adhesive products	19	label printing press	3
tableware	18	mail services	3
home food packaging products	17	airway cargo	2
cosmetics	15	newspaper subscription	2
groceries	14	rotorspinning	2
medical device	12	service contracts for elevators	2
respirators	12	welding equipment	2
electrodes	11	die casting machine	1
financial services	9	hammer drill	1
glassware	10	kitchen electronics	1
film for printing/graphics	8	ski rental	1
newspaper advertisement	8	train transportation	1
	Total = 435		

Appendix C: Selected Academic Studies on Price Elasticities – Further Information

Authors(s) (Year)	SKU/Brand	Data Source	Definition of Price	Sales vs. MS	Methodology	# of PE	Avg. PE.
Ailawadi/Gedenk/ Neslin (1999)	brand	IRI and Nielsen scanner panel data	actual price per ounce net of discounts and coupons for purchased brand, shelf price for other brands	MS	model comparison: model fit, forecast accuracy and marketing mix response	3 x 10 = 30	-1.87
Allenby (1989)	SKU	aggregated weekly, store level sales scanner data	price per pack (standardized to 4 rolls)	sales	random utility model to model discrete choice behavior	3 x 6 = 18	-4.34
Allenby/Rossi (1991)	SKU	household scanner panel data	actual price paid plus value of any redeemed coupons to establish shelf price	relative MS	model comparison (standard 1 price, standard 10 price, nested and nonhomothetic logit model)	4 x 10 = 40	-3.41
Bemmaor (1984)	SKU	not specified	unit price	MS	model comparison	4 x 5 = 20	-2.68
Besanko/Gupta/Jain (1998)	SKU	store level scanner data	retail shelf price per ounce	MS	model comparison	4 + 4 x 2 = 12	-2.71
Bolton (1989a)	brand	store level scanner data	not specified	sales	model comparison	(3 + 2 + 2 + 3) x 3 = 33 - 2 = 31	-2.40
Brodie/de Kluyver (1984)	brand	store level (retail audit) data	relative price	MS	model comparison	3 x 3 x 2 = 18	-0.78

Appendix 161

Authors(s) (Year)	SKU/Brand	Data Source	Definition of Price	Sales vs. MS	Methodology	# of PE	Avg. PE
Bucklin/Russell/ Srinivasan (1998)	brand	household scanner panel data	weighted avg. over the 4 available sizes	MS	multinomial logit choice model	9	-1.92
Carpenter et al. (1988)	brand	consumer panel	avg. price per month (monthly avg. of weekly store prices)	MS	asymmetric attraction model	11	-2.28
Chib/Seetharaman/ Strijnev (2004)	brand	scanner panel data	price per regular package size	MS	model comparison	4 x 2 = 8	-2.62
Chintagunta (1992)	SKU	scanner panel data	net purchase price if bought, otherwise shelf price	MS	nested logit model (model comparison)	4 x 2 = 8	-2.95
Chintagunta (1993)	SKU	scanner panel data	net purchase price	sales and MS	model comparison (proposed model vs. nested logit model)	4 x 3 = 12	-1.62
Chintagunta (2001)	SKU	sales (company data)	price of 16oz bottle	sales and MS	model comparison (probit and logit models)	(3 x 2 x 6) + 3 x 1 = 39	-1.21
Chintagunta/ Honore (1996)	brand	scanner panel data	net purchase price if bought, otherwise shelf price	MS	model comparison	4 x 4 = 16	-1.35
Chintagunta/Jain/ Vilcassim (1991)	SKU	scanner panel data	net purchase price	MS	semiparametric random effects model	4	-1.99
Christen et al. (1997)	SKU	market and store level scanner data	not specified	sales	comparison of estimation methods	5 x 4 + 2 x 2 x 2 = 28	-2.41
Cooper (1988)	brand	store level scanner data	price per pound	MS	three mode factor analysis	12	-1.78
Gönül/Srinivasan (1993)	brand	household scanner data	price per diaper	MS	model comparison (multinominal logit models)	3 x 4 = 12	-2.03

Authors(s) (Year)	SKU/Brand	Data Source	Definition of Price	Sales vs. MS	Methodology	# of PE	Avg. PE
Guadagni/Little (1983)	SKU	store and panel data	regular price, promotional price	MS	logit model	8 x 2 = 16	-2.24
Gupta et al. (1996)	SKU	household and store scanner data	price per ounce	MS	model comparison	10 x 6 + 7 x 3 = 81	-2.05
Hildebrandt/ Klapper (2001)	brand	store level scanner data	price and regular price divided by lowest price	MS	TUCK-AL S3	9	-2.86
Kadiyali/ Chintagunta/ Vilcassim (2000)	brand	retail scanner data (aggregated across stores)	retail price weighted across UPCs and sizes	sales in ounces	game theory based model	3 + 3 = 6	-3.26
Kalyanam (1996)	SKU	store level scanner data	retail price per pound	MS	Bayesian mixture model	6 x 2 = 12	-5.66
Kamakura/Russell (1989)	brand	household scanner data	price per ounce	sales in ounces	probabilistic choice model	4	-3.81
Kim (1995)	SKU	household and store level scanner data	price per SKU	MS	model comparison	3 x 6 = 18	-4.41
Kim/Allenby/Rossi (2002)	SKU	household scanner panel data	not specified (assumption all flavors have the same price)	sales	model comparison	5 x 4 = 20	-2.50
Kim/Rossi (1994)	SKU	household scanner panel data	price per SKU	MS	random coefficient logit model	5 x 2 = 10	-3.30
Kopalle/Mela/ Marsh (1999)	brand	Store level data	regular and mean price per ounce	sales	descriptive dynamic brand sales model	6	-1.63

Appendix

Authors(s) (Year)	SKU/Brand	Data Source	Definition of Price	Sales vs. MS	Methodology	# of PE	Avg. PE
Krishnamurti/Raj (1991)	brand	household scanner panel data	weighted avg. price per unit, coffee: shelf price/oz minus deal amount/oz	sales and MS	combination of logit and regression model	(3+3) x 2 = 12	-4.43
Kumar/Divakar (1999)	SKU	store level scanner data	avg. price per unit	MS	Rotterdam model	11 + 16 x 2 = 43	-0.20
Mantrala et al. (2006)	SKU	store level company data	avg. weekly price	sales	multinominal logit model	66	-2.22
Mehta/Rajiv/ Srinivasan (2003)	brand	household panel	price/oz	MS	model comparison	4 x 2 = 8	-1.07
Montgomery (1997)	SKU	store level scanner data	avg. weekly price standardized to 64oz	MS	Bayesian hierarchical model	11	-2.92
Moon/Russell/ Duvvuri (2006)	brand	household panel	avg. price per 1000 sheets	MS	different random utility models	5 x 3 x 2 = 30	-4.50
Mulhern/Williams/ Leone (1998)	SKU	store level scanner data	price per bottle	MS	2-stage econometric model	14	-3.52
Murthi/Srinivasan (1999)	SKU	household scanner data	price per ounce	MS	model comparison	6 x 2 = 12	-2.96
Reibstein/Gatignon (1984)	SKU	Store level data	price per pack	sales	model comparison	5 x 3 – 1 = 14	-1.68
Roy/Chintagunta/ Halder (1996)	SKU	household scanner data	cents per ounce	MS	model comparison	4 x 3 = 12	-3.09
Russell/Bolton (1988)	brand	store level scanner data	avg. price per pack	sales	model comparison	13 x 3 = 39	-3.09
Russell/Kamakura (1994)	brand	store level scanner data	cents per ounce (net of discounts)	MS	model comparison	10 x 3 = 30	-2.13

Authors(s) (Year)	SKU/Brand	Data Source	Definition of Price	Sales vs. MS	Methodology	# of PE	Avg. PE
Sivakumar (2001)	brand	scanner panel data	paid/faced price per 32 ounces	MS	model comparison	8	-4.21
Srinivasan/ Popkowski-Leszczyc/Bass (2000)	brand	store level (beer), household (marg.) scanner data	avg. weekly price per oz	MS	vector auto-regressive model	5 + 4 + 4 + 3 = 16	-1.21
Van Heerde/ Gupta/Wittink (2003)	SKU	household panel data	not specified	sales	proposed model	4 + 2 + 2 = 8	-3.65
Van Heerde/ Mela/Manchanda (2004)	brand	store level scanner data	avg. price per pound	MS	dynamic linear model, log-log model	5 x 2 +2 = 12	-2.80
Villas-Boas/Winer (1999)	SKU	household scanner panel data	avg. price per pack	MS	model comparison (random utility models)	(3 + 3) x 2 = 12	-5.74
Villas-Boas/Zhao (2005)	SKU	household scanner panel data	avg. price per pack	MS	model comparison	3 x 2 = 6	-3.29

Appendix D: Expert Interviews

In order to better understand the determinants of price elasticity, interviews with pricing experts in various industries were conducted. This lead to a quantitative analysis of 12 interviews which results are displays in figure Appendix D-1 to Appendix D-6. During the expert interviews product characteristics, market characteristics and customer characteristics were evaluated.

The respondents were asked about their level of agreement to the hypothesized effect of a determinant on the magnitude of price elasticity. The level of agreement was measured on a scale from 1 = "strongly disagree" to 5 = "strongly agree". In addition, the respondents were asked about the strength of the effect ranging from 1 = "no effect" to 5 = "very strong effect". The results of the assessment are displayed in figure D-1 for the product characteristics, figure D-3 for the market characteristics and figure D-5 for the customer characteristics.

In addition, the data is, as suggested by pricing expert Prof. Dr. Dr. h.c. mult. Hermann Simon, graphically displayed indicating the agreement of experts with the hypothesized effect. The answers were categorized into "agreement", "neither agreement nor disagreement", "disagreement" and "no answer". The results show the percentage of respondents for each category. The graphical illustrations of the results are displayed in figure D-2 for the product characteristics, figure D-4 for the market characteristics and figure D-6 for the customer characteristics.

Appendix D-1: Expert Interviews - Evaluation of Product Characteristics, Hypothesized Effect on Price Elasticity, Level of Agreement and Expected Strength of Effect

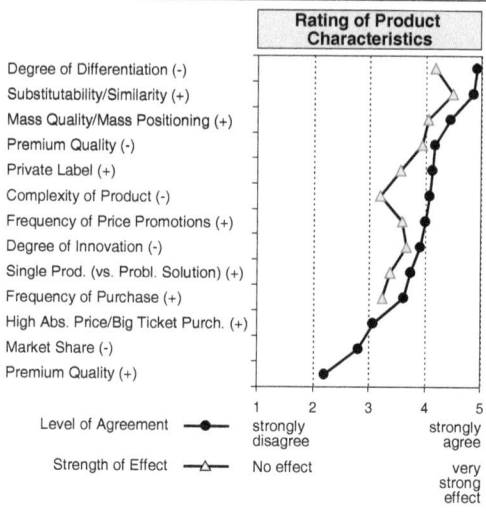

-: negative impact on magnitude of PE
+: positive impact on magnitude of PE
n = 12

Appendix D-2: Expert Interviews - Evaluation of Product Characteristics, Agreement with Hypothesized Effect on Price Elasticity

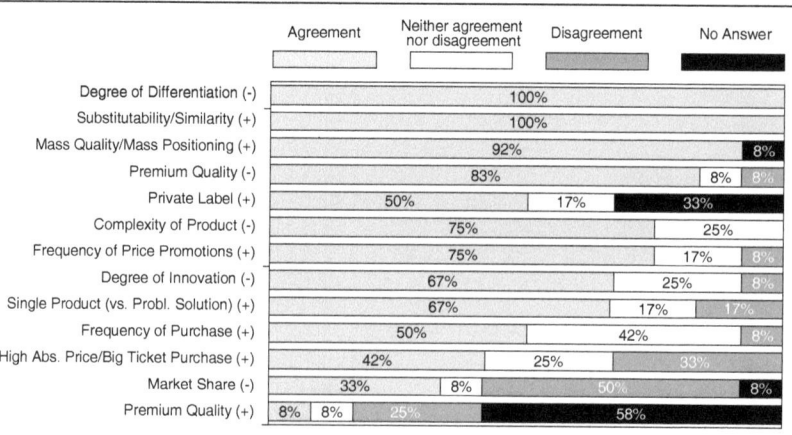

% of respondents

n = 12

Appendix D-3: Expert Interviews - Evaluation of Market Characteristics, Hypothesized Effect on Price, Level of Agreement and Expected Strength of Effect

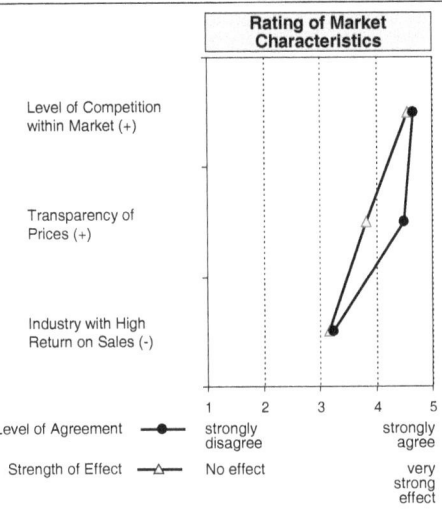

-: negative impact on magnitude of PE
+: positive impact on magnitude of PE
n = 12

Appendix D-4: Expert Interviews - Evaluation of Market Characteristics, Agreement with Hypothesized Effect on Price Elasticity

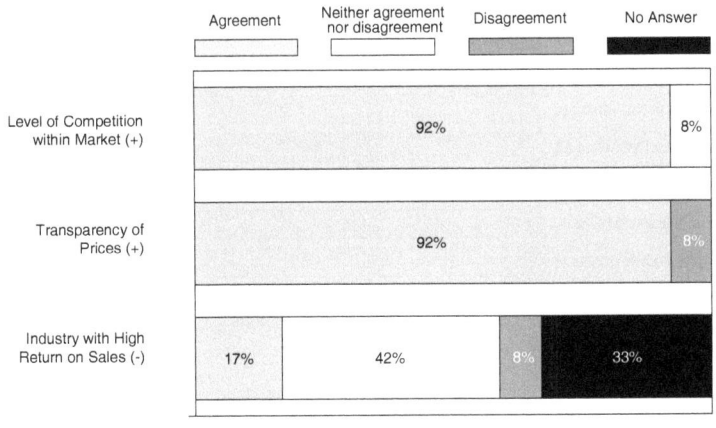

n = 12

Appendix D-5: Expert Interviews - Evaluation of Customer Characteristics, Hypothesized Effect on Price Elasticity, Level of Agreement and Expected Strength of Effect

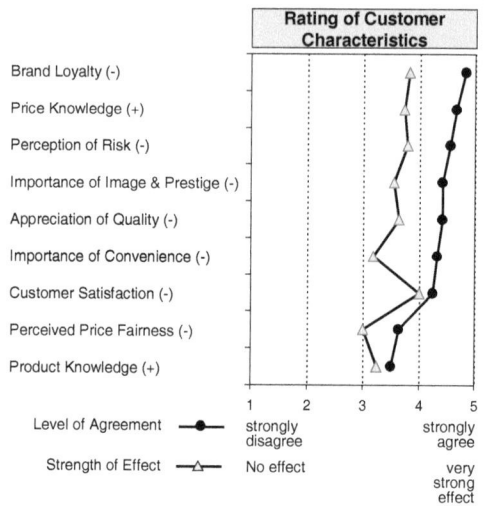

-: negative impact on magnitude of PE
+: positive impact on magnitude of PE
n = 12

Appendix D-6: Expert Interviews - Evaluation of Customer Characteristics, Agreement with Hypothesized Effect on Price Elasticity

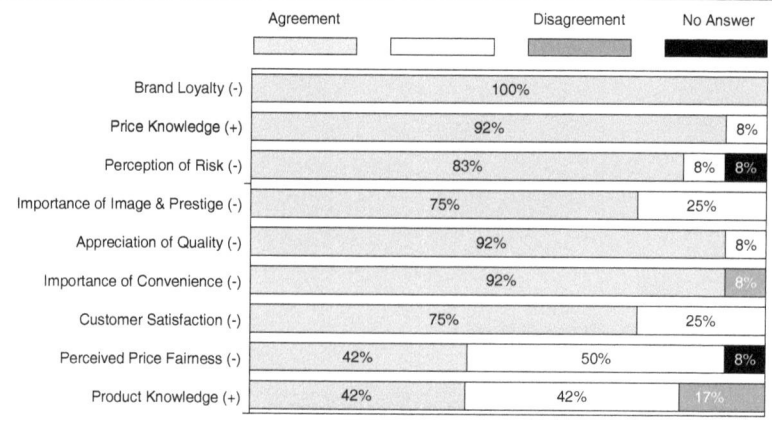

% of respondents
n = 12

References

Aaker, D.A./Keller, K.L. (1990): Consumer Evaluations of Brand Extensions, in: Journal of Marketing, Vol. 54, No. 1, pp. 27-41.

Ailawadi, K.L./Gedenk, K./Lutzky, C./Neslin, S.A. (2007): Decomposition of the Sales Impact of Promotion-Induced Stockpiling, in: Journal of Marketing Research, Vol. 44, No. 3, pp. 450-467.

Ailawadi, K.L./Gedenk, K./Neslin, S.A. (1999): Heterogeneity and Purchase Event Feedback in Choice Models: An Empirical Analysis with Implications for Model Building, in: International Journal of Research in Marketing, Vol. 16, No. 3, pp. 177-198.

Ailawadi, K. L./Harlam, B. (2004): An Empirical Analysis of the Determinants of Retail Margins: The Role of Store-Brand Share, in: Journal of Marketing, Vol. 68, No. 1, pp. 147-165.

Ailawadi, K.L./Lehmann, D.R./Neslin S.A. (2001): Market Response to a Major Policy Change in the Marketing Mix: Learnings from Procter & Gamble's Value Pricing Strategy, in: Journal of Marketing, Vol. 65, No. 1, pp. 44-61.

Ailawadi, K.L./Neslin, S.A./Gedenk, K. (2001): Pursuing the Value-Conscious Consumer: Store Brands Versus National Brand Promotions, in: Journal of Marketing, Vol. 65, No. 1, pp. 71-89.

Albers, S./Mantrala, M.K./Sridhar, S. (2008): A Meta-analysis of Personal Selling Elasticities, Marketing Science Institute Working Paper Series, No. 1, The Marketing Science Institute: Cambridge.

Albers, S./Mantrala, M.K./Sridhar, S. (2010): Personal Selling Elasticities: A Meta-analysis, in: Journal of Marketing Research, Vol. 47, No. 5, pp. 840-853.

Allenby, G.M. (1989): A Unified Approach to Identifying, Estimating and Testing Demand Structures with Aggregate Scanner Data, in: Marketing Science, Vol. 8, No. 3, pp. 265-280.

Allenby, G.M./Rossi, P.E. (1991): Quality Perceptions and Asymmetric Switching between Brands, in: Marketing Science, Vol. 10, No. 3, pp. 185-204.

Anderson, E.T./Song, I. (2004): Coordinating Price Reductions and Coupon Events, in: Journal of Marketing Research, Vol. 41, No. 4, pp. 441-422.

Ariely, D./Huber, J./Wertenbroch, K. (2005): When Do Losses Loom Larger Than Gains?, in: Journal of Marketing Research, Vol. 42, No. 2, pp. 134-138.

Backhaus, K./Erichson, B./Plinke, W./Weiber, R. (2006): Multivariate Analysemethoden: Eine anwendungsorientierte Einführung, 11th edition, Berlin, Heidelberg, New York: Springer.

Baker, W.L. /Marn, M.V./Zawada, C.C. (2010): The Price Advantage, 2nd edition, Hoboken, New Jersey: John Wiley & Sons.

Bass, F.M./Wind, J. (1995): Introduction to the Special Issue: Empirical Generalizations in Marketing, in: Marketing Science, Vol. 14, No. 3, pp. 1-5.

Bauerly, R./Johnson, D. T. (2005): An Evaluation of Journals Used in Doctoral Programs, in: Journal of the Academy of Marketing Science, Vol. 33, No. 3, pp. 313-329.

Baumgartner, H./Pieters, R. (2003): The Structural Influence of Marketing Journals: A Citation Analysis of the Discipline and Its Subareas Over Time, in: Journal of Marketing, Vol. 67, No. 2, pp. 123-139.

Bell, D.R./Chiang, J./Padmanabhan, V. (1999): The Decomposition of Promotional Response: An Empirical Generalization, in: Marketing Science, Vol. 18, No. 4, pp. 504-526.

Bell, D./Lattin, J. (2000): Looking for Loss Aversion in Scanner Panel Data: The Confounding Effect of Price Response Heterogeneity, in: Marketing Science, Vol. 19, No. 2, pp. 185-200.

Bemmaor, A.C. (1984): Testing Alternative Econometric Models on the Existence of Advertising Threshold Effect, in: Journal of Marketing Research, Vol. 21, No. 3, pp. 298-308.

Ben-Akiva, M./Bradley, M./Morikawa, T./Benjamin, J./Novak, T./Oppewal, H./Rao, V. (1994): Combining Revealed and Stated Preferences Data, in: Marketing Letters, Vol. 5, No. 4, pp. 335-350.

Bendle, N./Farris, P./Pfeifer, P./Reibstein, D. (2010): Metrics that Matter – to Marketing Managers, in: Marketing ZFP – Journal of Research and Management, Vol. 32, No 1, pp. 18-23.

Besanko, D./Gupta, S./Jain, D. (1998): Logit Demand Estimation Under Competitive Pricing Behavior: An Equilibrium Framework, in: Marketing Science, Vol. 44, No. 11, pp. 1533-1547.

Bidwell, M.O./Wang, B.X./ Zona, J.D. (1995): An Analysis of Asymmetric Demand Response to Price Changes: The Case of Local Telephone Calls, in: Journal of Regulatory Economics, Vol. 8, No. 3, pp. 285-298.

Bijmolt, T.H.A./Van Heerde, H.J./Pieters, R.G.M. (2005): New Empirical Generalizations on the Determinants of Price Elasticity, in: Journal of Marketing Research, Vol. 42, No. 2, pp. 141-156.

Blair, E./Zinkhan, G.M. (2006): Nonresponse and Generalizability in Academic Research, in: Journal of the Academy of Marketing Science, Vol. 34, No. 1, pp. 4-7.

Blankmeyer, E. (2006): How Robust is Linear Regression with Dummy Variables?, Faculty Publications-Finance and Economics, Department of Finance and Economics, Texas State University, https://digital.library.txstate.edu/handle/10877/4105, [19.11.2012].

Blattberg, R.C./Wisniewski, K.J. (1989): Price-Induced Patterns of Competition, in: Marketing Science, Vol. 8, No. 4, pp. 291-309.

Bolton, R. (1989a): The Robustness of Retail-Level Price Elasticity Estimates, in: Journal of Retailing, Vol. 65, No. 2, pp. 193-219.

Bolton, R. (1989b): The Relationship Between Marketing Characteristics and Promotional Price Elasticities, in: Marketing Science, Vol. 8, No. 2, pp. 153-169.

Boulding, W./Lee, E./Staelin, R. (1994): Mastering the Mix: Do Advertising, Promotion, and Sales Force Activities Lead to Differentiation? in: Journal of Marketing Research, Vol. 31, No. 2, pp. 159-172.

Brodie, R./de Kluyver, C.A. (1984): Attraction versus Linear and Multiplicative Market Share Models: An Empirical Evaluation, in: Journal of Marketing Research, Vol. 21, No. 2, pp. 194-201.

Brynjolfsson, E./Smith, M.D. (2000): Frictionless Commerce? A Comparison of Internet and Conventional Retailers, in: Management Science, Vol. 46, No. 4, pp. 563-585.

Bucklin, R.E./Russell, G.J./Srinivasan, V. (1998): A Relationship Between Market Share Elasticities and Brand Switching Probabilities, in: Journal of Marketing Research, Vol. 35, No. 1, pp. 99-113.

Bucklin, R.E./Srinivasan, V. (1991): Determining Interbrand Substitutability through Survey Measurement of Consumer Preference Structures, in: Journal of Marketing Research, Vol. 28, No. 1, pp. 58-71.

Burnham, T.A./Frels, J.K./Mahajan, V. (2003): Consumer Switching Costs: A Typology, Antecedents, and Consequences, in: Journal of the Academy of Marketing Science, Vol. 31, No. 2, pp. 109-126.

Carpenter, G.S./Cooper, L.G./Hanssens, D.M./Midgley, D.F. (1988): Modeling Asymmetric Competition, in: Marketing Science, Vol. 7, No. 4, pp. 393-412.

Carpenter, G.S./Glazer, R./Nakamoto, K. (1994): Meaningful Brands From Meaningless Differentiation: The Dependence on Irrelevant Attributes, in: Journal of Marketing Research, Vol. 31, No. 3, pp. 339-350.

Chandran, S./ Morwitz, V.G. (2006): The Price of "Free"-dom: Consumer Sensitivity to Promotions with Negative Contextual Influences, in: Journal of Consumer Research, Vol. 33, No. 3, pp. 384-392.

Che, H./Chen, X./Chen, Y. (2012): Investigating Effects of Out-of-Stock on Consumer Stockkeeping Unit Choice, in: Journal of Marketing Research, Vol. 49, No. 4, pp. 502-513.

Chen, H./Marmorstein, H./Tsiros, M./Rao, A.R. (2012): When More Is Less: The Impact of Base Value Neglect on Consumer Preferences for Bonus Packs over Price Discounts, in: Journal of Marketing, Vol. 76, No. 4, pp. 64-77.

Chib, S./Seetharaman, P.B./Strinijnev, A. (2004): Model of Brand Choice with a No-Purchase Option Calibrated to Scanner-Panel Data, in: Journal of Marketing Research, Vol. 41, No. 2, pp. 184-196.

Chintagunta, P.K. (1992): Heterogeneity in Nested Logit Models: An Estimation Approach and Empirical Results, in: International Journal of Research in Marketing, Vol. 9, No. 2, pp. 161-175.

Chintagunta, P.K. (1993): Investigating Purchase Incidence, Brand Choice and Purchase Quantity Decisions of Households, in: Marketing Science, Vol. 12, No. 2, pp. 184-208.

Chintagunta, P.K. (2001): Endogeneity and Heterogeneity in a Probit Demand Model: Estimation Using Aggregate Data, in: Marketing Science, Vol. 20, No. 4, pp. 442-456.

Chintagunta, P.K./Honore, B.E. (1996): Investigating the Effects of Marketing Variables and Unobserved Heterogeneity in a Multinomial Logit Model, in: International Journal of Research in Marketing, Vol. 13, No. 1, pp. 1-15.

Chintagunta, P.K./Jain, D.C./Vilcassim, N.J. (1991): Investigating Heterogeneity in Brand Preferences in Logit Models for Panel Data, in: Journal of Marketing Research, Vol. 28, No. 4, pp. 417-428.

Christen, M./Gupta, S./Porter, J.C./Staelin, R./Wittink, D.R. (1997): Using Market-Level Data to Understand Promotion Effects in a Nonlinear Model, in: Journal of Marketing Research, Vol. 34, No. 3, pp. 322-334.

Chu, J./Chintagunta, P./Cebollada, J. (2008): Research Note - A Comparison of Within-Household Price Sensitivity Across Online and Offline Channels, in: Marketing Science, Vol. 27, No. 2, pp. 283-299.

References

Cohen, A. (1991): Dummy Variables in Stepwise Regression, in: The American Statistician, Vol. 45, No. 3, pp. 226-228.

Cooper, H. (1998): Synthesizing Research: A Guide for Literature Reviews, Thousand Oak, CA: Sage.

Cooper, L.G. (1988): Competitive Maps: The Structure Underlying Asymmetric Elasticities, in: Management Science, Vol. 34, No. 6, pp. 707-723.

Danaher, P.J./Brodie, R.J. (2000): Understanding the Characteristics of Price Elasticities for Frequently Purchased Packaged Goods, in: Journal of Marketing Management, Vol. 16, No. 8, pp. 917-936.

Dickson, P.R./Ginter, J.L. (1987): Market Segmentation, Product Differentiation, and Marketing Strategy, in: Journal of Marketing, Vol. 51, No. 2, pp. 1-10.

Diller, H. (1997): Preismanagement im Zeichen des Beziehungsmarketing, in: Die Betriebswirtschaft, Vol. 57 No. 6, pp. 749-63.

Diller, H. (2004): Preisstrategien im Industriegütermarketing, in: Backhaus, K./Voeth, M. (eds.): Handbuch Industriegütermarketing: Strategien – Instrumente – Anwendungen, Gabler: Wiesbaden, pp. 947-968.

Diller, H. (2008): Preispolitik, 4th edition, Stuttgart: Kohlhammer.

Dolan, R.J. (1995): How Do You Know When the Price is Right?, in: Harvard Business Review, Vol. 73, No. 5, pp. 174-183.

Du, R.Y./Kamakura, W.A. (2006): Household Life Cycles and Lifestyles in the United States, in: Journal of Marketing Research, Vol. 43, No. 1, pp. 121-132.

Du, R.Y./Kamakura, W.A. (2008): Where Did All That Money Go? Understanding How Consumers Allocate Their Consumption Budget, in: Journal of Marketing, Vol. 72, No. 6, pp. 109-131.

Ehrenberg, A.S.C. (1995): Empirical Generalizations, Theory and Method, in: Marketing Science, Vol. 14, No. 3, pp. 20-28.

Estelami, H./Lehmann, D.R./ Holden, A.C. (2001): Macro-economic Determinants of Consumer Price Knowledge: A Meta-Analysis of Four Decades of Research, in: International Journal of Research in Marketing, Vol. 18, No. 4, pp. 341-355.

Ewing, J. (2004): Hidden Champions - The little-known European Companies that are Conquering the World, in: Business Week, January 26, pp. 42-44.

Fok, D./Horváth, C./Paap, R./Franses, P.H. (2006): A Hierachical Bayes Error Correction Model to Explain Dynamic Effects of Price Changes, in: Journal of Marketing Research, Vol. 43, No. 3, pp. 443-461.

Frank, R.H. (2008): Microeconomics and Behavior, 7th edition, Boston: Irwin McGraw-Hill.

Gabor, A. (1988): Pricing – Concepts and Methods for Effective Marketing, Hants: Gower.

Gavious, A./Lowengart, O. (2012): Price–Quality Relationship in the Presence of Asymmetric Dynamic Reference Quality Effects, in: Marketing Letters, Vol. 23, No. 1, pp. 137-161.

Ghosh, A./Neslin, S.A./Shoemaker, R.W. (1983): Are There Associations Between Price Elasticity and Brand Characteristics?, in: Murphy, P.E./Laczniak, E.R. (eds): AMA Educators' Conference Proceeding,. Chicago: American Marketing Association, pp. 226-230.

Ghosh, A./Neslin, S.A./Shoemaker, R.W. (1984): A Comparison of Market Share Models and Estimation Procedures, in: Journal of Marketing Research, Vol. 21, No. 2, pp. 202-210.

Gijsbrechts, E. (1993): Prices and Pricing Research in Consumer Marketing: Some Recent Developments, in: International Journal of Research in Marketing, Vol. 10, No. 2, pp. 115-151.

Glass G.V. (1976): Primary, Secondary, and Meta-analysis of Research, in: Educational Researcher, Vol. 5, No. 10, pp. 3-8.

Glass, G.V./McGaw, B./Smith, M.L. (1981): Meta-Analysis in Social Research, Newbury Park: Sage Publications.

Gönül, F./Srinivasan, K. (1993): Modeling Multiple Sources of Heterogeneity in Multinominal Logit Models: Methodological and Managerial Issues, in: Marketing Science, Vol. 12, No. 3, pp. 213-229.

Goldenberg, J./Horowitz, R./Levav, A./Mazursky, D. (2003): Finding your Innovation Sweet Spot, Harvard Business Review, Vol. 81, No.3, pp.120-129.

Goldsmith, R.E./Flynn, L.R./Goldsmith, E.B. (2003): Innovative Consumers and Market Mavens, in: Journal of Marketing Theory and Practice, Vol. 11, No. 4, pp. 54-65.

Goldsmith, R.E./Kim, D./Flynn, L.R./Kim, W.-M. (2005): Price Sensitivity and Innovativeness for Fashion Among Korean Consumers, in: The Journal of Social Psychology, Vol. 145, No. 5, pp. 501-508.

Gordon, B./Goldfarb, A./Li, Y. (2013): Does Price Elasticity Vary with Economic Growth? A Cross-category Analysis, in: Journal of Marketing Research, Vol. 50, No. 1, pp. 4-23.

Guadagni, P.M./Little, J.D. (1983): A Logit Model of Brand Choice Calibrated on Scanner Data, in: Marketing Science, Vol. 2, No. 3, pp. 203-238.

Gupta, S./Chintagunta, P.K./Kaul, A./Wittink, D.R. (1996): Do Household Scanner Data Provide Representative Inferences from Brand Choices: A Comparison with Store Data, in: Journal of Marketing Research, Vol. 33, No. 4, pp. 383-398.

Gurumurthy, K. and John D. C. Little (1989): A Price Response Model Developed From Perception Theories, Working Paper, Sloan School of Management: MIT.

Ha-Brookshire, J.E./Norum, P.S. (2011): Willingness to Pay for Socially Responsible Products: Case of Cotton Apparel, in: Journal of Consumer Marketing, Vol. 28, No. 5, pp. 344-353.

Hamilton, W./East, R./Kalafatis, S. (1997): The Measurement and Utility of Brand Price Elasticities, in: Journal of Marketing Management, Vol. 13, No. 4, pp. 285-298.

Hamilton, R.W./Srivastava, J. (2008): When 2 + 2 Is Not the Same as 1 + 3: Variations in Price Sensitivity Across Components of Partitioned Prices, in: Journal of Marketing Research, Vol. 45, No. 4, pp. 450-461.

Han, S./Gupta, S./Lehmann, D.R. (2001): Consumer Price Sensitivity and Price Thresholds, in: Journal of Retailing, Vol. 77, No. 4, pp. 435-456.

Hanssens, D./Parson, L.J./Schultz, R.L. (2001): Market Response Models: Econometric and Time Series Analysis, 2nd edition, Boston: Kluwer Academic Publishers.

Heil, O.P./Helsen, K. (2001): Toward an Understanding of Price Wars: Their Nature and How They Erupt, in: International Journal of Research in Marketing, Vol. 18, No. 1-2, pp. 83-98.

Hennig-Thurau, T./Houston, M.B./Heitjans, T. (2009): Conceptualizing and Measuring the Monetary Value of Brand Extensions: The Case of Motion Pictures, in: Journal of Marketing, Vol. 73, No. 6, pp. 167-183.

Hensel-Börner, S./Sattler, H. (2000): Ein empirischer Validitätsvergleich zwischen der Customized Computerized Conjoint Analysis (CCC), der Adaptive Conjoint Analysis (ACA) und Self-Explicated-Verfahren, in: Zeitschrift für Betriebswirtschaft, Vol. 70, No. 6, pp. 705-725.

Herrmann, A. (2003): Relevanz des Preismanagements für den Unternehmenserfolg, in: Diller, H./Herrmann, A. (editors): Handbuch Preispolitik: Strategien – Planung – Organisation – Umsetzung, Wiesbaden: Gabler, pp. 33-48.

Herrmann, A./Huber, F./Sivakumar, K./Wricke, M. (2004): An Empirical Analysis of the Determinants of Price Tolerance, in: Psychology & Marketing, Vol. 21, No. 7, pp. 533-551.

Herrmann, A./Xia, L./Monroe, K./Huber, F. (2007): The Influence of Price Fairness on Customer Satisfaction, in: Journal of Product & Brand Management, Vol. 16, No.1, pp. 49-58.

Hildebrandt, L./Klapper, D. (2001): The Analysis of Price Competition Between Corporate Brands, in: International Journal of Research in Marketing, Vol. 18, No. 1-2, pp. 139-159.

Hoch, S.J./Kim, B./Montgomery, A.L./Rossi, P.E. (1995): Determinants of Store-Level Price Elasticity, in: Journal of Marketing Research, Vol. 32, No. 1, pp. 17-29.

Homburg, C./Koschate, N. (2005a): Behavioral Pricing-Forschung im Überblick, Teil 1: Grundlagen, Preisinformationsaufnahme und Preisinformationsbeurteilung, in: Zeitschrift für Betriebswirtschaft, Vol. 75, No. 4, pp. 383-423.

Homburg, C./Koschate, N. (2005b): Behavioral Pricing-Forschung im Überblick, Teil 2: Preisinformationsspeicherung, weitere Themenfelder und zukünftige Forschungsrichtungen, in: Zeitschrift für Betriebswirtschaft, Vol. 75, No. 5, pp. 501-524.

Huang, M.-H./Jones, E./Hahn, D.E. (2007): Determinants of price elasticities for private labels and national brands of cheese, in: Applied Economics, Vol. 39, No. 5, pp. 553-563.

Hult, G.T./Neese, W.T./Bashaw, R.E. (1997) Faculty Perceptions of Marketing Journals, in: Journal of Marketing Education, Vol. 19, No. 1, pp. 37-52.

Hunter, J.E./Schmidt, F.L. (2004): Methods of Meta-Analysis: Correcting Error and Bias in Research Findings, 2nd edition, Thousand Oaks, CA: Sage.

Jedidi, K./Mela, C.F./Gupta, S. (1999): Managing Advertising and Promotion for Long-Run Profitability, in: Marketing Science, Vol. 18, No. 1, pp. 1-22.

Jensen, R.T./Miller, N.H. (2008): Giffen Behavior and Subsistence Consumption, in: The American Economic Review, Vol. 98, No. 4, pp. 1553-1577.

Johnston, W.J./Lewin, J.E. (1996): Organizational Buying Behavior: Toward an Integrative Framework, in: Journal of Business Research, Vol. 35 No.1, pp.1-15.

Kachersky, L. (2011): Reduce Content or Raise Price? The Impact of Persuasion Knowledge and Unit Price Increase Tactics on Retailer and Product Brand Attitudes, in: Journal of Retailing, Vol. 87, No. 4, pp. 479–488.

Kadiyali, V./Chintagunta, P./Vilcassim, N. (2000): Manufacturer-Retailer Channel Interactions and Implications for Channel Power: An Empirical Investigation of Pricing in a Local Market, in: Marketing Science, Vol. 19, No. 2, pp. 127-148.

Kahneman, D./Tversky, A. (1979): Prospect Theory: An Analysis of Decision under Risk, in: Econometrica, Vol. 47, No. 2, pp. 263-291.

Kalra, A. /Goodstein, R.C. (1998): The Impact of Advertising Positioning Strategies on Consumer Price Sensitivity, in: Journal of Marketing Research, Vol. 35, No. 2, pp. 210-224.

Kalwani, M.U./Yim, C.K. (1992): Consumer Price and Promotion Expectations: An Experimental Study, in: Journal of Marketing Research, Vol. 29, No. 1, pp. 90-100.

Kalyanam, K. (1996): Pricing Decisions Under Demand Uncertainty: A Bayesian Mixture Model Approach, in: Marketing Science, Vol. 15, No. 3, pp. 297-221.

Kamakura, W.A./Russell, G.J. (1989): A Probabilistic Choice Model for Market Segmentation and Elasticity Structure, in: Journal of Marketing Research, Vol. 26, No. 4, pp. 379-390.

Kaul, A./Wittink, D. (1995): Empirical Generalizations about the Impact of Advertising on Price Sensitivity and Price, in: Marketing Science, Vol. 14, No. 3, pp. 151-160.

Kim, B.-D. (1995): Incorporating Heterogeneity with Store-Level Aggregate Data, in: Marketing Letters, Vol. 2, No. 2, pp. 159-169.

Kim, B.-D./Rossi, P.E. (1994): Purchase Frequency, Sample Selection, and Price Sensitivity The Heavy User Bias, in: Marketing Letters, Vol. 5, No. 1, pp. 57-67.

Kim, B.-D./Srinivasan, K./Wilcox, R.T. (1999): Identifying Price Sensitive Consumers: The Relative Merits of Demographic vs. Purchase Pattern Information, in: Journal of Retailing, Vol. 75, No. 2, pp. 173-193.

Kim, J./Allenby, G.M./Rossi, P.E. (2002): Modeling Consumer Demand for Variety, in: Marketing Science, Vol. 21, No. 3, pp. 229-250.

Klein, B./Leffler, K.B. (1981): The Role of Market Forces in Assuring Contractual Performance, in: Journal of Political Economy, Vol. 89, No. 4, pp. 615-641

Kopalle, P.K./Mela, C.F./Marsh, L. (1999): The Dynamic Effect of Discounting on Sales: Empirical Analysis and Normative Pricing Implications, in: Marketing Science, Vol. 18, No. 3, pp. 317-332.

Koschate, N. (2002): Kundenzufriedenheit und Preisverhalten – Theoretische und empirisch experimentelle Analysen, Wiesbaden: Gabler.

Krishnamurthi, L./Raj, S.P. (1991): An Empirical Analysis of the Relationship between Brand Loyalty and Consumer Price Elasticity, in: Marketing Science, Vol. 10, No. 2, pp. 172-183.

Kucher, E. (1985): Scannerdaten und Preissensitivität bei Konsumgütern, Wiesbaden: Gabler.

Kumar, P./Divakar, S. (1999): Size Does Matter: Analyzing Brand-Size Competition Using Store Level Scanner Data, in: Journal of Retailing, Vol. 75, No. 1, pp. 59-76.

Lauszus, D./Ebel, B. (2000): Preisabsatzfunktionen, in: Herrmann, A./Homburg, C. (eds.): Marktforschung: Methoden, Anwendungen, Praxisbeispiele, 2nd edition, Wiesbaden: Gabler, pp. 833-859.

Lee, H.L./Tang, C.S. (1997): Modelling the Cost and Benefit of Delayed Product Differentiation, in: Management Science, Vol. 43, No. 1, pp. 40-53.

Leeflang, P.S.H./Wittink, D.R. (2000): Building Models for Marketing Decisions: Past, Present and Future, in: International Journal of Research in Marketing, Vol. 17, No. 2-3, pp. 105-126.

Leeflang, P.S./Wieringa, J.E. (2010): Modeling the Effects of Pharmaceutical Marketing, in: Marketing Letters, Vol. 21, No. 2, pp. 121-133.

Leone, R.P./Robinson, L.M./Bragge, J./Somervuori, O. (2012): A citation and profiling analysis of pricing research from 1980 to 2010, in: Journal of Business Research, Vol. 65, No. 7, pp. 1010–1024.

Lichtenstein, D.R./Ridgway, N.M./Netemeyer, R.G. (1993): Price Perceptions and Consumer Shopping Behavior: A Field Study, in: Journal of Marketing Research Vol. 30, No. 2, pp. 234-245.

Lipsey, M.W./Wilson, D.B. (2001): Practical Meta-analysis, Thousand Oaks: Sage Publications.

Lodish, L./Mela, C.F. (2008): Manage Brands over Years, Not Quarters, in: The Pricing Advisor, Marietta: Professional Pricing Society, pp. 5-7.

Lu, Q./Moorthy, K.S. (2007): Coupons vs. Rebates, in: Marketing Science, Vol. 26, No. 1, pp. 67-82.

Lynch, J.G./Ariely, D. (2000): Wine Online: Search Costs Affect Competition on Price, Quality, and Distribution, in: Marketing Science, Vol. 19, No. 1, pp. 83-103.

Madzumdar, T./Raj, S.P./Sinha, I. (2005): Reference Price Research: Review and Propositions, in: Journal of Marketing, Vol. 69, No. 4, pp. 84-102.

Mantrala, M.K./Seetharam, P.B./Kaul, R./Gopalakrishna, S./Stam, A. (2006): Optimal Pricing Strategies for an Automotive Aftermarket Retailer, in: Journal of Marketing Research, Vol. 43, No. 4, pp. 588-604.

Marshall, A. (1890): Principles of Economics, 1st edition, London: Macmillan.

Marshall, A. (1895): Principles of Economics, 3rd edition, London: Macmillan.

Matzler, K./Würtele, A./Renzl, B. (2006): Dimensions of Price Satisfaction: A Study in the Retail Banking Industry, in: International Journal of Bank Marketing, Vol. 24, No. 4, pp. 216-231.

McQuiston, D.H. (1989): Novelty, Complexity, and Importance as Casual Determinants of Industrial Buyer Behavior, in: Journal of Marketing, Vol. 53, No. 2, pp. 66-79.

Mehta, N./Rajiv, S./Srinivasan, K. (2003): Price Uncertainty and Consumer Search: A Structural Model of Consideration Set Formation, in: Marketing Science, Vol. 22, No. 1, pp. 58-84.

Miller, K.M./Hofstetter, R./Krohmer, H./Zhang, Z.J. (2011): How Should Consumers' Willingness to Pay be Measured? An Empirical Comparison of State-of-the-Art Approaches, in: Journal of Marketing Research, Vol. 48, No. 1, pp. 172-184.

Mitra, A./Lynch Jr, J.G. (1995): Toward a Reconciliation of Market Power and Information Theories of Advertising Effects on Price Elasticity, in: Journal of Consumer Research, Vol. 21, No. 4, pp. 644-659.

Mizik, N./Jacobson, R. (2008): The Financial Value Impact of Perceptual Brand Attributes, in: Journal of Marketing Research, Vol. 45, No. 1, pp. 15-32.

Monroe, K.B. (2003): Pricing: Making Profitable Decisions, 3rd edition, Boston: Mc Graw Hill.

Montgomery, A.L. (1997): Creating Micro-Marketing Pricing Strategies Using Supermarket Scanner Data, in: Marketing Science, Vol. 16, No. 4, pp. 315-337.

Moon, S./Russell, G.J./Duvvuri, S.D. (2006): Profiling the Reference Price Consumer, in: Journal of Retailing, Vol. 82, No. 1, pp. 1-11.

Moosmayer, D.C./Wendlandt, M./Patz, O. (2009), Determinanten und Verhaltensfolgen eines erhöhten Preisinteresses, in: Marketing ZFP, Vol. 31, No. 3, pp. 159-170.

Mukherjee, A./Hoyer, W.D. (2001): The Effect of Novel Attributes on Product Evaluation, in: Journal of Consumer Research, Vol. 23, No. 3, pp. 462-472.

Mulhern, F.J./Williams, J.D./Leone, R.P. (1998): Variability of Brand Price Elasticities across Retail Stores: Ethnic, Income, and Brand Determinants, in: Journal of Retailing, Vol. 74, No. 3, pp. 427-446.

Murthi, B.P./Srinivasan, K. (1999): Consumers' Extent of Evaluation in Brand Choice, in: Journal of Business, Vol. 72, No. 2, pp. 229-256.

N.N. (2010): Deutsche Raststätten sind zu teuer und zu schlecht, July 6, http://www.welt.de/motor/verkehr/article8317436/Deutsche-Raststaetten-sind-zu-teuer-und-zu-schlecht.html [21.07.2011].

N.N. (2011a): Raststätten-Kunden klagen über hohe Preise, June 30, http://www.focus.de/panorama/welt/verkehr-raststaetten-kunden-klagen-ueber-hohe-preise_aid_641621.html [21.07.2011].

N.N. (2011b): Vielen Deutschen sind Autobahnraststätten zu teuer, June 30, http://www.welt.de/reise/nah/article13460584/Vielen-Deutschen-sind-Autobahnraststaetten-zu-teuer.html [21.07.2011].

Narasimhan, C./Neslin, S.A./Sen, S.K. (1996): Promotional Elasticities and Category Characteristics, in: Journal of Marketing, Vol. 60, No. 2, pp. 17-30.

Narayanan, S./Desiraju, R./Chintagunta, P.K. (2004): Return on Investment Implications for Pharmaceutical Promotional Expenditures: The Role of Marketing-Mix Interactions, in: Journal of Marketing, Vol. 68, No. 4, pp. 90-105.

Naudé, P./Desai, J./Murphy, J. (2003): Identifying the Determinants of Internal Marketing Orientation, in: European Journal of Marketing, Vol. 37, No. 9, pp. 1205-1220.

Novemsky, N./Kahneman, D. (2005): The Boundaries of Loss Aversion, in: Journal of Marketing Research, Vol. 42, No. 2, pp. 119-128.

Nowlis, S.M./Simonson, I. (1996): The Effect of New Product Features on Brand Choice, in: Journal of Marketing Research, Vol. 33, No. 1, pp. 36-46.

Parker, P.M./Gatignon, H. (1996): Oder of Entry, Trial Diffusion, and Elasticity, Dynamics: An Empirical Case, in: Marketing Letters, Vol. 7, No. 1, pp. 95-109.

Parker, P.M./Neelamegham, R. (1997): Price Elasticity Dynamics over the Product Life Cycle: A Case Study of Consumer Durables, in: Marketing Letter, Vol. 8, No. 2, pp. 205-208.

Pauwels, K./Srinivasan, S./ Franses, P.H. (2007): When Do Price Thresholds Matter in Retail Categories?, in: Marketing Science, Vol. 21, No. 1, pp. 83-100.

Porter, M.E. (1985): Technology and Competitive Advantage, in: Journal of Business Strategy, Vol. 5, No. 3, pp.60 - 78

Putler, D.S. (1992): Incorporating Reference Price Effects into a Theory of Consumer Choice, in: Marketing Science, Vol. 11, No. 3, pp. 287-309.

Raju, J.S. (1992): The Effect of Price Promotions on Variability in Product Category Sales, in: Marketing Science, Vol. 11, No. 3, pp. 207-220.

Raman, K./Bass, F.M. (1988): A General Test of Reference Price Theory in the Presence of Threshold Effects, Working Paper, College of Business Administration, University of Florida.

Ramirez, E./Goldsmith, R. (2009): Some Antecedents of Price Sensitivity, in: Journal of Marketing Theory and Practice, Vol. 17, No. 3, pp. 199-213.

Reibstein, D.J./Gatignon, H. (1984): Optimal Product Line Pricing: The Influence of Elasticities and Cross-Elasticities, in: Journal of Marketing Research, Vol. 21, No. 3, pp. 259-267.

Rogers, E.M. (1995): Diffusion of Innovations, New York: Free Press.

Rosenthal, R./Rubin, D.B. (1978): Interpersonal Expectancy Effects: The First 345 Studies, in: The Behavioral and Brain Sciences, Vol. 1, No. 3, pp. 377-415.

Roth, S. (2010): Preismanagement – Stand der aktuellen Lehrbuch- und Managementliteratur (Sammelrezension), in: Die Betriebswirtschaft, Vol. 70, No. 2, pp. 165-187.

Roy, R./Chintagunta, P.K./Haldar, S. (1996): A Framework for Investigating Habits, "The Hands of the Past", and Heterogeneity in Dynamic Brand Choice, in: Marketing Science, Vol. 15. No. 3, pp. 280-299.

Russell, G.J./Bolton, R.N. (1988): Implications of Market Structure for Elasticity Structure, in: Journal of Marketing Research, Vol. 25, No. 3, pp. 229-241.

Russell, G. J./Kamakura, W.A. (1994): Understanding Brand Competition Using Micro and Macro Scanner Data, Vol. 31, No. 2, pp. 289-303.

Schlegelmilch, B.B./Bohlen, G.M./Diamantopoulos, A. (1996): The link between green purchasing decisions and measures of environmental consciousness, in: European Journal of Marketing, Vol. 30, No. 5, pp. 35-55.

Schlosser, A.E./White, T.B./Lloyd, S.M. (2006): Converting Web Site Visitors into Buyers: How Web Site Investment Increases Consumer Trusting Beliefs and Online Purchase Intentions, in: Journal of Marketing, Vol. 70, No. 2, pp. 133-148.

Schmidt, F.L./Hunter, J.E. (1977): Development of a General Solution to the Problem of Validity Generalization, in: Journal of Applied Psychology, Vol. 62, No. 5, pp. 529-540.

Sethuraman, R./Tellis, G.J. (1991): An Analysis of the Tradeoff Between Advertising and Price Discounting, in: Journal of Marketing Research, Vol. 28, No. 2, pp. 160-174.

Sethuraman, R./Tellis, G.J./Briesch, R.A. (2011): How Well Does Advertising Work? Generalizations from Meta-Analysis of Brand Advertising Elasticities, in: Journal of Marketing Research, Vol. 48, No. 3, pp. 457-471.

Shocker, A.D./Bayus, B.L./Kim, N. (2004): Product Complements and Substitutes in the Real World The Relevance of the "Other Products", in: Journal of Marketing, Vol. 68, No. 1, pp. 28-40.

Shostack, G.L. (1977): Breaking Free from Product Marketing, in: Journal of Marketing, Vol. 41, No. 2, pp. 73-80.

Silva, A./Nayga Jr., R.M./Campbell, B.L./Park, J. (2007): On the Use of Valuation Mechanisms to Measure Consumers' Willingness to Pay for Novel Products: A Comparison of Hypothetical and Non-Hypothetical Values, in: International Food and Agribusiness Management Review, Vol. 10, No. 2, pp. 165-180.

Simon, H. (1979): Dynamics of Price Elasticity and Brand Life Cycles: An Empirical Study, in: Journal of Marketing Research, Vol. 16, No. 4, pp. 439-452.

Simon, H. (1989): Price Management, Amsterdam-New York: Elsevier.

Simon, H./Bilstein, F./Luby, F. (2006): Manage for Profit, Not for Market Share: A Guide to Greater Profits in Highly Contested Markets, Massachusetts: Harvard Business School Press.

Simon, H./Dolan, R.J. (1997): Profit durch power pricing: Strategien aktiver Preispolitik, Frankfurt: Campus.

Simon, H./Fassnacht, M. (2009): Preismanagement: Strategie – Analyse – Entscheidung – Umsetzung, 3rd edition, Wiesbaden: Gabler.

Sivakumar, K. (2001): Measuring Consumer Response to Price Using Logit Models: Implications of Ignoring Category Purchase Aspect, in: Journal of Marketing Theory and Practice, Vol. 9, No. 2, pp. 1-10.

Slater, S.F./Narver, J.C. (2000): The Positive Effect of a Market Orientation on Business Profitability: a Balanced Replication, in: Journal of Business Research, Vol. 48, No. 1, pp. 69-73.

Smith, M.D./Brynjolfsson (2001): Consumer Decision Making at an Internet Shopbot: Brand Still Matters, Journal of Industrial Economics, Vol. 49, No. 4, pp. 541-558.

Smith, M.L./Glass, G.V. (1977): Meta-analysis of Psychotherapy Outcome Studies, in: American Psychologist, Vol. 32, No. 9, pp. 752-760.

Song, I./Chintagunta, P.K. (2007): A Discrete-Continuous Model for Multicategory Purchase Behavior of Households, in: Journal of Marketing Research, Vol. 44, No. 4, pp. 595-612.

Srinivasan, S./Popkowski-Leszczyc, P.T.L./Bass, F.M. (2000): Market Share Response and Competitive Interaction: The Impact of Temporary, Evolving and Structural Changes in Price, in: International Journal of Research in Marketing, Vol. 17, No. 4, pp. 281-305.

Stamer, H./Liebermann, H.P. (2004): Untersuchung des Einflusses von Determinanten des Entscheidungskontexts beim Produktkauf auf das Preisinteresse in Konsumgütermärkten, Working Paper No. 123, Nürnberg: Working Papers of the Chair of Marketing at the University Erlangen Nürnberg.

Stigler, G.J. (1947): Notes on the History of the Giffen Paradox, in: Journal of Political Economy, Vol. 55, No. 2, pp.152–156.

Swaminathan, V. (2003): The Impact of Recommendation Agents on Consumer Evaluation and Choice: The Moderating Role of Category Risk, Product Complexity, and Consumer Knowledge, in: Journal of Consumer Psychology, Vol. 13, No. 1/2, pp. 93-101.

Tellis, G.J. (1988): The Price Elasticity of Selective Demand: A Meta-Analysis of Econometric Models of Sales, in: Journal of Marketing Research, Vol. 25, No. 4, pp. 331-341.

Tellis, G.J./Fornell, D.C. (1988): The Relationship Between Advertising and Quality over the Product Life Cycle. A Contingency Theory, in: Journal of Marketing Research, Vol. 15, No. 1, pp. 64-71.

Thaler, R. (1985): Mental Accounting and Consumer Choice, in: Marketing Science, Vol. 4, No. 3, pp. 199-214.

Theoharakis, V./Hirst, A. (2002): Perceptual Differences of Marketing Journals: A Worldwide Perspective, in: Marketing Letters, Vol. 13, No. 4, pp. 389-402.

Toppin, G./Czerniawska, F. (2005): Business Consulting. A Guide to How it Works and How to Make it Work, edition „The Economist", London: Economist Books.

Twardawa, W. (2010): Ökonomische und psychologische Auswirkungen der Rezession auf das Konsumentenverhalten, Presentation of GfK Consumer Tracking, Panel Services Deutschland (October 28).

Van Heerde, H.J./Gijsbrechts, E./Pauwels, K. (2008): Winners and Losers in a Major Price War, in: Journal of Marketing Research, Vol. 45, No. 5, pp. 499-518.

Van Heerde, H.J./Gupta, S./Wittink, D.R. (2003): Is 75% of the Sales Promotion Bump Due to Brand Switching? No, Only 33% Is, in: Journal of Marketing Research, Vol. 40, No. 4, pp. 481-491.

Van Heerde, H.J./Mela, C.F./Manchanda, P. (2004): The Dynamic Effect of Innovation on Market Structure, in: Journal of Marketing Research, Vol. 41, No. 2, pp. 166-183.

Veblen, T. (1899): The Theory of the Leisure Class, New York: Macmillan.

Veisten, K. (2007): Willingness to Pay for Eco-labelled Wood Furniture: Choice-based Conjoint Analysis versus Open-ended Contingent Valuation, in: Journal of Forest Economics, Vol. 13, No. 1, pp. 29-48.

Villas-Boas, J.M./Winer, R.S. (1999): Endogeneity in Brand Choice Models, in: Management Science, Vol. 45, No. 10, pp. 1324-1338.

Villas-Boas, J.M./Zhao, Y. (2005): Retailer, Manufacturers, and Individual Consumers: Modeling the Supply Side in the Ketchup Marketplace, in: Journal of Marketing Research, Vol. 42, No. 1, pp. 83-95.

Völckner, F. (2006a): Methoden zur Messung individueller Zahlungsbereitschaften: ein Überblick zum State of the Art, in: Journal für Betriebswirtschaft, Vol. 56, No. 1, pp. 33-60.

Voelckner, F (2006b): An Empirical Comparison of Methods for Measuring Consumers' Willingness to Pay, in: Marketing Letters, Vol. 17, No. 2, pp. 137-49.

Völckner, F./Hofmann, J. (2007): The Price-perceived Quality Relationship: A Meta-analytic Review and Assessment of Its Determinants, in: Marketing Letters, Vol. 18, No. 3, pp. 181-196.

Wagner, U./Taudes, A. (1991): Microdynamics of a New Product Purchase: A Model Incorporating both Marketing and Consumer-specific Variables, in: International Journal of Research in Marketing, Vol. 8, No. 3, pp. 223-249.

Webster, F.E./Wind, Y. (1972): A General Model for Understanding Organization Buying Behavior, in: Journal of Marketing, Vol. 36, No. 2, pp. 12-19.

Weiber, R./Rosendahl, T. (1997): Anwendungsprobleme der Conjoint-Analyse: Die Eignung conjointanalytischer Untersuchungsansätze zur Abbildung realer Entscheidungsprozesse, in: Marketing – Zeitschrift für Forschung und Praxis, Vol. 19, No. 2, pp. 107-118.

Whittingham, M.J./Stephens, P.A./Bradbury, R.B./Freckleton, R.P. (2006): Why Do We Still Use Stepwise Modelling in Ecology and Behaviour?, in: Journal of Animal Ecology, Vol. 75, No. 5, pp. 1182–1189.

Woodside, A.G. (2012): Consumer Evaluations of Competing Brands: Perceptual versus Predictive Validity, in: Psychology & Marketing, Vol. 29, No. 6, pp. 458-466.

Wricke, M. (2000): Preistoleranz von Nachfragern, Wiesbaden: Gabler.

Wricke, M./Herrmann, A./Huber, F. (2000): Behavioral Pricing – Erklärung- und Operationalisierungsansätze des Referenzpreiskonzepts, in: WiSt, Vol. 29, No. 12, pp. 692-697.

Xia, L./Monroe, K.B./Cox, J.L. (2004): The Price is Unfair! A Conceptual Framework of Price Fairness Perceptions, in: Journal of Marketing, Vol. 68, No. 4, pp. 1-15.

Yau, O.H./McFetridge, P.R./Chow, R.P./Lee, J.S./Sin, L.Y./Alan, C.B. (2000): Is Relationship Marketing for Everyone?, in: European Journal of Marketing, Vol. 34, No. 9/10, pp. 1111-1127.

Zeithaml, V.A. (1988): Consumer Perceptions of Price, Quality and Value: A Means-End Model and Synthesis of Evidence, in: Journal of Marketing, Vol. 52, No. 3, pp. 2-22.

Zeithaml, V.A./Berry, L.L./Parasuraman, A. (1996): The Behavioral Consequences of Service Quality, in: Journal of Marketing, Vol. 60, No. 2, pp. 31-46.

Schriften zu Marketing und Handel

Herausgegeben von Martin Fassnacht

Band 1 Ibrahim Köse: Qualität elektronischer Dienstleistungen: Messung und Auswirkungen. 2007.

Band 2 Christina Reith: Convenience im Handel. 2007.

Band 3 Christina Müller: Differenzierung von Handelsunternehmen. 2007.

Band 4 Sonja Klose: Gefährdung existierender Kundenbeziehungen. 2008.

Band 5 Saskia Hardwig: Der Einfluss des Geschäftsstättenimages auf die Produktbewertung von Handelsmarken. 2008.

Band 6 Andreas Ettinger: Auswirkungen von Einkaufsconvenience. 2010.

Band 7 Jochen Mahadevan: Wahrgenommene Preisfairness bei personenbezogener Preisdifferenzierung. 2010.

Band 8 Christian M. Wiegner: Preis-Leistungs-Positionierung. Konzeption und Umsetzung. 2010.

Band 9 Frank Breitschwerdt: Preismanagement von Konsumgüterherstellern. Konzeption, Umsetzung und Erfolgsauswirkungen. 2011.

Band 10 Christian Stallkamp: Betriebsformen im Automobilhandel. Konzeptualisierung und empirische Ergebnisse einer multiattributiven Präferenzstrukturmodellierung. 2011.

Band 11 Katia Rumpf: Preis und Markendehnung. Eine empirische Analyse. 2011.

Band 12 Henning Mohr: Der Preismanagement-Prozess für Luxusmarken. Gestaltung und Erfolgsauswirkungen. 2013.

Band 13 Yorck Nelius: Organisation des Preismanagements von Konsumgüterherstellern. Eine empirische Untersuchung. 2013.

Band 14 Rebecca Winkelmann: Preisdifferenzierung aus Kundensicht. Eine verhaltenswissenschaftliche Untersuchung wahrgenommener Komplexität von Preisdifferenzierung. 2013.

Band 15 Sabine El Husseini: EDLP versus Hi-Lo Pricing Strategies in Retailing. Literature Review and Empirical Examinations in the German Retail Market. 2014.

Band 16 Evelyn Friedel: Price Elasticity. Research on Magnitude and Determinants. 2014.

www.peterlang.com

www.ingramcontent.com/pod-product-compliance
Ingram Content Group UK Ltd.
Pitfield, Milton Keynes, MK11 3LW, UK
UKHW041902230426
12049UKWH00002B/13